SIGNALS, NOISE, AND ACTIVE SENSORS

SIGNALS, NOISE, AND ACTIVE SENSORS

Radar, Sonar, Laser Radar

JOHN MINKOFF
Distinguished Member of the Technical Staff
AT&T Bell Laboratories
Whippany, New Jersey

A Wiley-Interscience Publication
JOHN WILEY & SONS, INC.
New York • Chichester • Brisbane • Toronto • Singapore

Library of Congress Cataloging in Publication Data:

Minkoff, John.

Signals, noise, and active sensors: radar, sonar, laser radar / John Minkoff.

p. cm.

"A Wiley-Interscience publication."

Includes bibliographical references.

1. Signal processing--Mathematics. 2. System analysis. 3. Radar. 4. Sonar. I. Title.

TK5102.5.M54 1991

621.382'2--dc20 91-16515

ISBN 0-471-54572-4 CIP

Printed in the United States of America

10 9 8 7 6 5 4 3 2 1

CONTENTS

Preface ix

1. Introduction—Fundamentals of Receivers 1

2. Review of Probability 13

2.1 Bernoulli Trials—The Binomial Distribution, 13
2.2 The Poisson Distribution, 15
2.3 The Exponential Distribution, 16
2.4 The Gaussian Distribution, 16
2.5 The Rayleigh and Rice Distributions, 16
2.6 Joint Distributions, Conditional Distributions, and Bayes Theorem, 17
2.7 Characteristic Functions, 18
2.8 The Law of Large Numbers, 20
2.9 The Central Limit Theorem, 21
2.10 Approximations to the Gaussian Distribution, 23
2.11 Functions of a Random Variable, 25

3. Review of Noise and Random Processes 28

3.1 Introduction—Correlation Functions and Power Spectral Densities, 28
3.2 Types of Noise, 31
3.3 Power Spectral Density of Thermal Noise—Nyquist's Theorem, 33
3.4 Power Spectral Density of Shot Noise, 35
3.5 Shot Noise and Optical Receivers—The Quantum Limit, 39

3.6 Noise Statistics—Shot Noise in Radar and Laser Radar, 42
3.7 Noise Figure and Noise Temperature, 48
3.8 Noise Figure of an Attenuator, 50
3.9 Applications: Noise Power Measurements, 51
3.10 Connections with Statistical Physics, 55

4. Continuous and Discrete-Time Signals **61**

4.1 The Sampling Theorem and Oversampling, 61
4.1.1 Application of the Sampling Theorem to Delay of Discrete–Time Signals, 69
4.2 The Sampling Theorem for Bandpass Carrier Signals, 70
4.3 Signal Duration and Bandwidth, 72
4.4 The Analytic Signal, 74
4.5 Processing of Continuous and Discrete-Time Signals, 77

5. Detection of Signals in Noise **87**

5.1 Statistical Decision Theory: The Likelihood Ratio Test, 87
5.2 Decision Criteria—Bayes, Maximum Likelihood, Neyman–Pearson, 90
5.3 Implementation of Decision Criteria, 93
5.3.1 Gaussian Noise, 93
5.3.2 Shot Noise—Poisson Distribution, 96
5.4 Correlation Detection: The Matched Filter—I, 99
5.4.1 The Gaussian Channel, 99
5.4.2 The Shot Noise Channel, 106
5.5 The Matched Filter—II, 107

6. Coherent and Noncoherent Detection and Processing, **117**

6.1 Ideal Noncoherent Detection of a Single Pulse, 118
6.2 Comparison of Coherent and Noncoherent Detection, of a Single Pulse, 124
6.3 Improvement in Signal-to-Noise Ratio by Coherent and Noncoherent Integration, 129
6.3.1 Noncoherent Integration, 129
6.3.2 Coherent Integration, 135
6.4 Performance of Coherent and Noncoherent Integration, 139
6.4.1 Noncoherent Integration, 140
6.4.2 Coherent Integration, 142
6.5 Summary of Coherent and Noncoherent Detection and Processing, 144

7. Parameter Estimation and Applications **148**

7.1 Estimation of Range to a Target, 149
7.2 Generalized Parameter Estimation, 158
7.2.1 The Cramer–Rao Lower Bound on the Variance of an Estimator, 159
7.2.2 Maximum Likelihood Estimation, 161
7.3 Applications of Maximum Likelihood Estimation to Sensor Measurements, 162
7.3.1. Calculation of the Cramer–Rao Bound for Coherent and Noncoherent Observations, 164
7.4 Application of Parameter Estimation to Tracking and Prediction, 169

8. Waveform Analysis, Range-Doppler Resolution and Ambiguity **178**

8.1 Waveform Analysis, 179
8.2 Range-Doppler Resolution and Ambiguity—The Generalized Ambiguity Function, 183

9. Large Time-Bandwidth Waveforms **191**

9.1 Chirp Waveforms and Pulse Compression, 193
9.2 Doppler Invariant Properties of Chirp Waveforms, 198
9.3 Hyperbolic Frequency Modulation, 202
9.4 Ambiguity Function for Large *BT* Waveforms, 205
9.5 Coded Waveforms, 207

10. Generalized Coherent and Noncoherent Detection and Processing **212**

10.1 Noncoherent Detection of a Single Pulse, 213
10.2 Coherent and Noncoherent Integration, 215

11. Systems Considerations **218**

11.1 Beampatterns and Gain of Antennas and Arrays, 218
11.2 The Radar and Sonar Equations, 224
11.3 The Search Problem, 228
11.4 Specification of the False-Alarm Probability, P_{fa}, 229

Appendix Table of Values the Error Function **233**

References **239**

Index **241**

PREFACE

The invention of radar, which was stimulated by the events that eventually led to World War II and developed on a very large scale during the war, has had a profound influence on modern science and technology. Demands on radar performance led to development of a capability for generation, high-power transmission, and reception of wide-band electromagnetic pulses to a level far beyond what had previously been accomplished, or even considered to be a practical possibility. In parallel with these essentially hardware-oriented efforts, development of new mathematical concepts, analytical schemes and computational techniques also took place, which were necessitated by the problem of detection and extraction of information from the inevitably weak radar echoes in the presence of random interference. The famous work by N. Wiener (see References), on which techniques commonly employed today for dealing with noise in linear systems are based, was stimulated by the radar fire-control problem, and actually first appeared as a classified report. These efforts established much of the basis for the modern era of digital communications. These technologies, concepts and methods are now routinely applied not only in communication systems, but also in commercial and scientific applications of all kinds.

One of the earliest expositions of the problem of detection and extraction of information from—in this case—radar signals, was given in the unique work by Woodward (see References) which also discussed the connection between problems peculiar to radar and the fundamental Communications problem. At the present time, in addition to radar, sonar (which actually predates radar) and laser radar have developed to the point where many commercial and scientific, as well as military, applications exist for these sensing systems as well. Each of these types of sensors incorporates certain problems peculiar to themselves. Some of these arise from differences in the physical nature of the signals that are transmitted and in their propagation speeds. In each case however, the solutions to the problems associated with signal detection and acquisition of information are to be found by application of various subsets of Communication and Estimation theory, and it should therefore be possible to deal with these systems on a unified basis provided the aforementioned differences are taken into account, which is the purpose of this book.

This book has benefited significantly from numerous conversations with friends and colleagues both in and outside AT&T Bell Laboratories, whose management I also wish to thank for their encouragement and support during its preparation. The material is based on lecture notes for a course on this subject which I have given at Polytechnic University in New York. The material can be covered in a 3-hour one-semester course, and possibly also a 2-hour course with some selectivity. Also, the first six chapters by themselves are suitable for a one-semester course in signal processing, exclusive of sensor operations.

JOHN MINKOFF

Whippany, NJ

To Sue

With love and very best wishes.

John

SIGNALS, NOISE, AND ACTIVE SENSORS

1

INTRODUCTION—FUNDAMENTALS OF RECEIVERS

This book has been written for the individual with a background in mathematics, engineering or physics who wishes to become familiar with, or more knowledgeable about, the concepts and mathematical techniques that come into play in the design, analysis and use of active sensing systems for civilian, military and scientific purposes. By active sensing systems we mean systems that transmit a signal, with the intention that it will be scattered by an object of interest, and derive information about the object on the basis of information extracted from the received scattering; the scattered signal is sometimes referred to as the echo. In radar the signal is some form of electromagnetic wave in the radio-frequency range; hence the name, which is an acronym for Radio Detection and Ranging. In the case of sonar (Sound Detection and Ranging) the signal is an acoustic wave with frequencies generally ranging from tens to thousands of hertz transmitted in water, and in laser radar the signal is an electromagnetic wave at frequencies beginning at, and extending beyond, the infrared. The latter is sometimes therefore also referred to as lidar or ladar (Light or Laser Detection and Ranging) but we do not adopt this usage here. Active systems are in contrast to passive systems, which do not transmit a signal but effectively listen for sound waves or electromagnetic waves originating from a possible object of interest. Although passive systems are not dealt with explicitly here, many of the signal-processing techniques to be discussed are applicable to passive as well as active systems.

Acoustic systems need not necessarily be restricted to operations in water. For example, bats employ a form of sonar which is used in maneuvering around obstructions while in flight and also in locating prey. Also there are numerous medical applications which employ acoustic waves propagating in animal tissue at ultrasonic frequencies [1]—($\sim$1–10 MHz in these applications) some of which are essentially ultrasonic sonars used for diagnostic purposes. Their usefulness arises from the relatively low sound speeds (essentially that of sound in water $\sim$1500 m/s) yielding ultrasonic wavelengths of the order of 10^{-3}–10^{-4} m. This permits imaging of structures in the body having larger dimensions, which of course includes most organs. In other applications, rate of blood flow is measured nonintrusively by means of the Doppler shift imparted to the acoustic signal by the motion of the blood. These sensors employ the same signal processing techniques used in ordinary sonar, which are dealt with here. However the explicit treatment of acoustic-wave systems in what follows is for propagation in bodies of water.

Of the three categories of active systems under consideration the earliest to be used was sonar, whose development was initiated during World War I (WWI) for the purpose of detecting submarines under the name ASDIC (Allied Submarine Detection Investigation Committee). This work, which included the efforts of Lord Rutherford, did not actually bear fruit until after the war, in the early 1920s, when the first experiments with operational active submarine detection were carried out. The name sonar, an acronym in obvious parallel to radar, was adopted later in the American development of this technique during World War II (WWII).

The possibility that electromagnetic waves could be used to detect the presence of objects was recognized from the beginning of the development of radio early in the 20th century, and was actually proposed by Marconi in 1922 as a means to detect ships. During the 1920s American scientists at the U.S. Naval Research Laboratory proposed the use of an electromagnetic fence consisting of a continuous wave (CW) transmission across a body of water which, on being interrupted by the presence of a ship, would serve as a detector of ships. Also during this period the technology of transmitting short radio-frequency pulses and observing their echoes was developed for ionospheric research, primarily in England and in the United States. By the 1930s investigation of radio frequency scattering with a view towards development of practical detection systems was taking place in the United States and throughout Europe. British radar development, which was begun in the mid-1930s explicitly for purposes of air defense in response to the rapidly worsening political situation in Europe at the time, was evidently initiated by the Director of Scientific Research at the British Air Ministry as a result of his request that the Radio Research Establishment advise "on the practicality of proposals of the type colloquially called 'death ray' " [2]. The idea was that a sufficiently strong electromagnetic field could possibly be used to heat up an attacking aircraft to the extent that living tissue on-board would be destroyed and bombs exploded. That such a weapon in principle is possible has since been demonstrated with lasers, but at the time it was of course found to be not feasible. However, it was quickly recognized that if radio waves could not be used to destroy attacking aircraft, they could be used to detect their presence at long ranges, long before they are visible optically, at night, and under conditions of poor visibility. Development proceeded rapidly, owing in no small part to the experience gained by the development of the aforementioned pulse techniques for ionospheric work. Also, this effort was unique at the time in its integration of radar into an overall coordinated efficient air-defense system which, as is well known, was a key element in the victory by the British in the Battle of Britain.

Although radar development in the United States was initially not as advanced as in England, with the outbreak of WWII in Europe in 1939 the U.S. effort was intensified to the extent that mobile relatively high-performance radars were available by the time of American entry into the war two years later. This is evidenced by the fact that the Japanese aircraft that attacked Pearl Harbor were in fact detected and tracked by a mobile radar unit on the island of Oahu while they were still over 100 miles from their destination. Unfortunately, owing to the perceived unlikeliness of this event, the information provided by the radar was misinterpreted. We shall see in what follows that the a priori probability of an event can be an important quantitative factor in the interpretation of information provided by a sensor. Subsequent to United States' entry into the war, radar development was increased enormously, taking place primarily at the MIT Radiation Laboratory, of which the MIT Lincoln Laboratory is an outgrowth, and at the Bell Telephone Laboratories. In fact, much of the theory of signal detection presented herein was developed during WWII. Significant development also took place at the Service laboratories, which emphasized the lower frequency ranges (e.g. $\leq$800 Mhz); the MIT effort was concentrated mostly in the microwave range (roughly 10^3–10^4 MHz).

After WWII radar development continued, particularly in the United States, stimulated primarily by the cold-war environment. In 1946 it was demonstrated for the first time, at Ft. Monmouth NJ, that radar development had progressed to the point where detection of echoes reflected from the moon was possible, which has ultimately led to the science of radar astronomy. Another interesting development occurred around the end of the 1950s when it was discovered that radars operating in the High Frequency (HF)[†] range (~3–30 MHz, roughly corresponding to the short wave radio band) were capable of detecting and tracking targets at very long ranges (e.g. 2000 km) by means of reflection of transmissions, and reception of reflected target echoes, from the earth's ionosphere. This discovery, which was made and originally developed under great secrecy at the U.S. Naval Research Laboratory, has led to the technology of Over-the-Horizon

[†] Other nomenclature for radio/radar frequencies is: Very High Frequency (VHF) 30–300 MHz; Ultra High Frequency (UHF) 300–1000 MHz. The microwave range and above is divided into "bands," initially introduced during WWII for reasons of security. These are: L band 1000–2000 MHz, S band 2000–4000 MHz, C band 4000–8000 MHz, X band 8–12.5 GHz, Ku band 12.5–18 GHz, K band 18–26.5 GHz, Ka band 26.5–40 GHz, millimeter waves >40 GHz.

(OTH) radar, which is active today in military and in other applications. OTH signals scattered from the ocean surface can be used to provide information about wind velocities at remote locations, which can be useful in oceanography and in weather prediction. A very innovative effort in the 1950s, which employed techniques used in optical holography, led to the development of the side-looking or synthetic-aperture radar. In this type of radar a moving platform, such as an aircraft or a satellite, illuminates an area of land or water and, by means of appropriate signal-processing techniques, effectively synthesizes a very large antenna aperture which yields an angle resolution (see Chapter 11) orders of magnitude greater (better) than that which could be provided by the physical aperture actually employed.

The early interest in air defense of the continental United States, for which the first NIKE radar/missile systems were developed, was, with the demonstration in the late 1950s by the USSR of the feasibility of accurate targeting of Intercontinental Ballistic missiles (ICBM), expanded to include defense against ICBMs, and also submarine-launched missiles. This led to the development of the Ballistic Missile Early Warning System (BMEWS), which followed the earlier DEW (Distant Early Warning) line, and which consists of a network of sensors positioned at high latitudes for the purpose of detecting and tracking possible attacks originating from over the north pole in the Eastern hemisphere. OTH radars have also been included in this early warning system. In addition to these large, static, ground-based systems for strategic applications, significant developments were also made in airborne, shipboard and tactical air-defense radars, requiring development of low-cost, light-weight components.

Because of the extremely high speeds of ballistic missiles, ~7 km/s, and the overwhelming number of them that can be deployed, it becomes impossible for radars operated by humans in such applications to be effective. During this period therefore, which continued at a high level into the 1970s, there was a great development in digital computer-controlled sensor operations, particularly as applied to the use of large multifunction array antennas for simultaneous detection, tracking and prediction of the positions of large numbers of vehicles traveling at hypersonic speeds. In addition, there were significant developments in the design of sophisticated waveforms and signal processing techniques for detecting and tracking targets in the presence of junk produced by the breakup after re-entry into the earth's atmosphere of components of the spent rocket delivery systems. Also, scientific capabilities of sensing systems were significantly advanced by application of these techniques to gathering information concerning the physics of high-velocity bodies entering the atmosphere. There have been numerous commercial and scientific applications of these technologies, for example in the area of air traffic control, satellite-based remote sensing for monitoring weather conditions and earth resources, and so on. Possibly of greatest importance however, this research program has served to give a clearer picture of what is and what is not feasible, and possible, in the area of missile defense, which would otherwise be a matter of continuing speculation. This has no doubt been a contributing factor to the relative stability that has existed for some time with regard to these matters.

With the development of high-power lasers in the 1960s, it was of course immediately apparent that lasers could also have applications as signal sources in active sensing systems under conditions of good visibility. Laser radars are currently being used to measure spatial distributions and concentrations of atmospheric constituents, including pollutants. In one approach, these are determined from range-resolved measurements at two wavelengths, selected such that one is strongly absorbed and the other less so, if at all. This is known as the differential absorption lidar (DIAL) technique, which has been used to measure ozone and water-vapor profiles [3, 4]. The latter can also be determined from Raman scattering by H_2O molecules stimulated by laser illumination [5]. In all these cases it is the wavelength range required for this phenomenology—ranging from the infrared to the ultraviolet—which makes lasers suited for these applications. Also because of the very short wavelengths, very high angular-measurement accuracy can be achieved by relatively small laser systems with apertures of, say, 5–10 cm, which could be accomplished only with much greater difficulty at radar frequencies, for which much greater antenna apertures would be required. With this capability very accurate tracking and targeting of orbiting objects can be achieved by laser systems mounted in satellites or aircraft, which has been considered for military applications.

The MIT Radiation Laboratory series which documents the radar-development work carried out at MIT during WWII encompasses 28 volumes. It is therefore clear that the subject of active sensing systems covers a very wide range of science and technology. Even from the very beginning in the development of radar, however, it was recognized—as in fact it has long been recognized for any physical observation—that the amount of information that can be obtained from an observation is fundamentally limited by the random processes affecting it. Therefore, although technical areas such as power generation, transmission characteristics, electrical circuitry, and the like are essential elements of any sensing system, once these parameters are established the ultimate capability of the resulting system is determined by the waveforms and signal-detection schemes employed, and by the signal-processing techniques utilized to extract the desired information from the inevitably randomly fluctuating observables.

These random fluctuations can be separated into two categories. In propagating to and from the object of interest the transmitted and scattered signals can experience randomly varying propagation conditions. These can include random density gradients in the propagation medium, as well as reflections from mountains and other obstructions. When the target-echo propagation path includes reflections this can give rise to multipath, which in

TV transmission is manifested as ghost images. Although such reflections are actually deterministic, they are effectively random unless the nature of the terrain over which propagation takes place is known exactly. Multipath is a very important consideration in sonar, as are atmospheric density gradients for laser radar. Unwanted reflections in radar are generally termed clutter and these effects can be loosely categorized as arising from clutter in one form or another. In sonar, such reflections are termed reverberation, which, in addition to multipath, is a particularly severe problem because of the possibility of reflections from the ocean bottom and the surface, as well as scattering from shipping and marine life [6, 7].

Also, the scattering object itself will in general be moving and therefore vibrating, pitching, tumbling, and this motion imparts additional random fluctuations to the observables that arise in the scattering process. At radio frequencies these effects are referred to as scintillation, and in laser radar they are known as speckle. This phenomenon is particularly severe in laser radar since, because of the extremely short wavelengths, very small, even microscopic, movements of the scattering object can cause violent fluctuations in the strength of the scattered signal in any given direction, which can be a major limiting factor on target-tracking accuracy for laser radar [8]. Fluctuations in the observed signal amplitude, whether due to scintillation, speckle or propagation conditions, are commonly referred to as fading.

As indicated, these phenomena vary widely among the three systems under consideration, and are also peculiar to both the particular physical propagation conditions as well as the nature of the scattering object. Furthermore, depending on the situation they can sometimes be eliminated or reduced by appropriate measures. Perhaps more fundamental however are the random fluctuations that arise in the receiver path because of random motion of electrons in the various receiver system components. These random currents constitute noise, which is an inescapable phenomenon in any sensing system in which the effective receiver system temperature is above 0 K. And, whereas the subject of random fluctuations arising in propagation and scattering is highly specialized, requiring different treatments tailored to the particular system and the situation of interest, the more fundamental question regarding the means for dealing with the limitations imposed by noise can be treated generally, for all systems, if certain differences are accounted for. These differences arise because of physical differences in the transmitted signals. In particular, (1) the energy per photon in laser radar is greater than that of a radio-frequency photon by a factor of about 10^8, which leads to a type of noise in laser radar that is not present in radio or acoustic-wave systems, and (2) the speed c of signal propagation for radar and laser radar, 3×10^8 m/s, exceeds that of sound in water by a factor of about 10^5.

With regard to (1), for any scattered laser signal the number of photons incident on a detector per unit time is a random variable. Therefore, since optical receivers essentially count the number of photons that arrive during a given detection interval, optical signals incorporate a certain degree of essential randomness within themselves. One speaks of this randomness in the observable as being produced by shot noise. Thus, in laser radars, shot noise must also be considered in addition to the noise arising from random electron currents in receiver components. Shot noise, being an inescapable component of the signal, is fundamentally different from noise due to random currents, which is additive. Furthermore, shot noise can be the more important consideration in laser radars because, as will be seen, there are techniques for effectively eliminating noise due to random currents, and, in addition, photodetectors can be made to be extremely quiet. Thus the random photon arrival rate can be the fundamental limiting factor. Recently, it has been found that this limitation can also be somewhat reduced by the use of squeezed states [9], but we do not deal with this here. In radar the signal is of course also an electromagnetic wave, and the echo is therefore also a collection of photons. However, because the photon energy is so small, the number of photons arriving per second in a radar signal capable of being observed in the presence of noise due to random currents is, necessarily, extremely large, and the fluctuations in arrival times of individual photons become smoothed out and effectively eliminated. In radar, therefore, shot noise is not a consideration. In contrast, because the energy per optical photon is large, and other forms of noise may be very low, the photon arrival rate for the power levels employed in practical laser radar systems can be relatively low, and the fluctuations in photon arrival times can be very much in evidence. This is discussed in greater detail in Section 3.6.

With regard to (2), it will be shown that when signals scattered from a moving object are processed, the parameter v/c, where v is the range-rate (i.e. the component of target velocity along the sensor line of sight), is an important factor. For radar and laser radar, even for targets traveling at orbital escape velocities, ~7 km/s, the ratio v/c is very small, $\sim 2 \times 10^{-5}$. On the other hand, for sonar systems, even for slow targets, say with $v = 10$ knots, we have $v/c \sim 3 \times 10^{-3}$ which is orders of magnitude larger, and this can result in important differences in the means used to extract the desired information from acoustic vs. electromagnetic signals.

A second but less important difference arises because the fractional bandwidth—that is, the ratio of the signal bandwidth† to the nominal transmitted signal frequency—is usually, though not always, quite small in electromagnetic systems but most often rather large (i.e. close to unity) in acoustic systems.

To summarize, while a great many factors come into play in the design and operation of active sensors, once various system parameters such as transmitted signal power and power losses in the circuitry have been

† The frequency range within which most—say ~90%—of the signal energy is contained. There are a number of such definitions. This will be made more precise in what follows, as needed.

established, we are primarily concerned with what takes place in the receiver, specifically with regard to the extraction of the desired information from the scattered signal in the presence of randomly varying interference. This interference consists of essentially two components. One component arises outside the receiver in the propagation and scattering processes and is highly specific to the type of system, the particular propagating conditions and the nature of the target, as are the means for ameliorating its effects. The second component, noise, is more fundamental, and can never be entirely eliminated since it depends on the basic thermodynamic characteristics of the receiver components and the statistical laws governing photon emission. For this reason it is possible to deal with the noise problem in generality, for all types of active sensors, provided the aforementioned differences are taken into account. Furthermore, the statistical techniques developed herein for dealing with noise can in many cases be applied to dealing with clutter and the random processes arising in the scattering.

A diagram of a generic receiver which has typically been used in sensing systems is presented in Figure 1.1, together with the following descriptive paragraphs, for the purpose of introducing certain common, traditional, nomenclature and concepts. It should however be noted that considerable departures from this configuration can be found in present-day systems.

1. *Signal Collector and Transducer*. In a radar the antenna collects the scattered signal and converts the electromagnetic wave to currents that drive the front-end amplifier. In sonar the collector is a hydrophone or an array of hydrophones that convert the incident acoustic signals to electrical signals. In a laser radar the collector is the optical aperture. In radar and sonar there is the question of the gain of the antenna or the array which is discussed in Chapter 11.

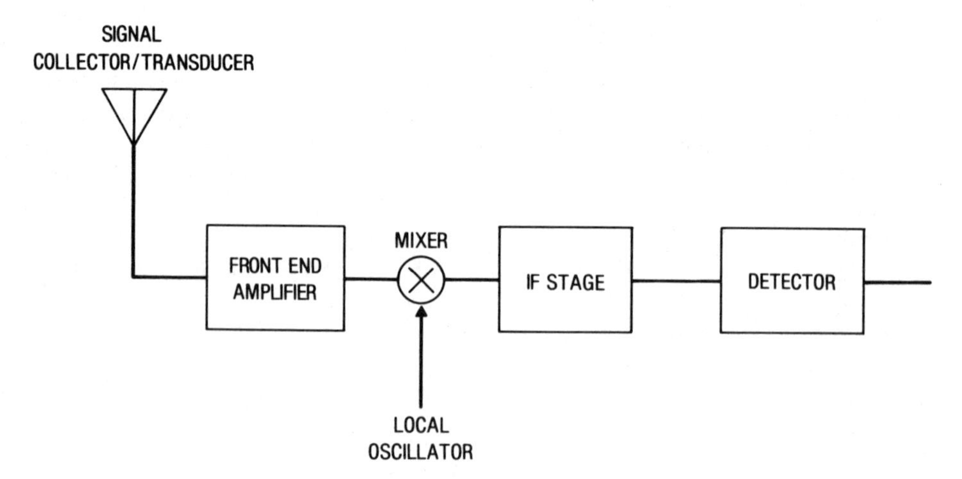

Figure 1.1 Generic receiver.

2. *Front-End Amplifier*. This stage of amplification is usually required because the received signal is, typically, too weak to overcome the noise of receiver components which follow. In laser radar systems if there is amplification it is achieved by means of photo multipliers.

3. *Local Oscillator and Mixer*. A typical signal at the output of the receiver front end might be of the form $a(t)\cos 2\pi f_0 t$, where the modulation $a(t)$ is a function which is slowly varying in time in comparison with f_0. The need for this stage arose originally in radar because, among other reasons, the radio frequency f_0 is generally too high for effective implementation of the subsequent receiver operations. The local-oscillator signal is of the form $\cos 2\pi f_{LO} t$ and multiplication of the front-end output would in this case yield $\frac{1}{2}a(t)\cos 2\pi f_{IF} t + \frac{1}{2}a(t)\cos 2\pi(f_O + f_{LO})t$.

The signal frequency f_O is thereby translated down to the more manageable intermediate frequency (IF), given by: $f_{IF} = f_O - f_{LO}$; the unwanted $f_O + f_{LO}$ term is rejected by the IF stage which follows. This frequency-translation process is known as heterodyning. Heterodyne receivers can also be used in laser radar, in which case the local oscillator is a coherent light source.

In sonar systems the signal frequency is not too high in the sense just described. However, heterodyning is sometimes employed with $f_{LO} = f_O$, in which case for the above example the mixer output is $\frac{1}{2}a(t) + \frac{1}{2}a(t)\cos 2\pi(2f_O)t$. In this way the signal amplitude $a(t)$, which in general may be complex, can be recovered directly. The advantages of this will be discussed in the relevant sections of the text.

4. *Intermediate Frequency Stage*. This stage may include further amplification of the frequency-translated output of the mixer if this is necessary. In all cases however the IF stage includes filtering to reject the $f_O + f_{LO}$ terms and also to reject noise outside the passband of interest, as defined by the signal bandwidth. This is necessary because the bandwidth of the front end can in general be wider than that of the signal. The IF stage is almost always present in radar receivers but will not be present in, nonheterodyne, sonar and laser radar receivers.

In all cases, although the signals are real, the observable quantity can be regarded as a time varying complex number with information both in its magnitude in its phase. If phase information in the signal is preserved in the process of observation the operation is said to be coherent. If the phase information in the signal is destroyed or ignored in the process of observation the operation is said to be noncoherent and only the magnitude of the signal is available.

5. *Detection Stage*. In the early days of radio a detector was implicitly a device with a nonlinear input/output characteristic—such as a rectifier, which passes only positive (or negative) portions of a signal—which stripped off the audio-frequency amplitude modulation of the transmitted radio-frequency (RF) carrier. Actually, almost any nonlinear device will do this,

as will other things not designed for the purpose, such as junctions between components of bed springs, between fillings and teeth, radiator valve connections, etc. which may allow electrical current to flow across the junction in one direction only. The crystal radio set operates on this principle, and, in fact, in the early days, before the Federal Communications Commission set limits on broadcaster's transmitted power, reports of voices emanating unaccountably from mattresses, people's teeth, wall units, etc. were not uncommon, because the unamplified modulation of the RF signal, detected in this manner, was often sufficiently powerful to produce an audible sound wave.

This nomenclature was carried over into the early development of radar, in which the detector was either a linear rectifier (envelope detector) or a square-law device, which yielded either the magnitude or the magnitude squared of the complex number. Such devices are still in use today, but the detection process has become much more sophisticated and can include linear phase-preserving operations as well as nonlinear ones. Also, whereas all signal processing operations were originally performed by circuits consisting of resistors, capacitors and inductors operating on continuous voltages and currents, today, sensor signals are most often sampled, quantized to certain discrete predetermined levels by analog-to-digital converters, encoded to a sequence of, say, ones and zeros, and all the necessary processing steps are carried out on the transformed discrete-time signal by a computer, often especially designed for the purpose. Nevertheless, the basic function of what is commonly referred to as Detection has not changed. It is to extract the desired information from the, possibly modified, received signal and present the information to the user or to the observer.

The final output of the receiver was originally termed the video signal[†] in the early days of radar development because it is at this point that the information conveyed by the signal can be made available for visual observation by a human observer in the form of a trace on an oscilloscope. In present-day systems the observer is of course often a computer rather than a human. The usage however has persisted, and the output of the receiver is commonly referred to as the video signal, whether it is actually converted to a visual image or not. In communication systems the receiver output is commonly referred to as the baseband signal. Both these terms are used interchangeably throughout the text. The mathematical definition of a video or baseband signal is that its spectrum (i.e. its Fourier transform) includes DC (i.e. the frequency origin, $f = 0$). If the Detection operation is nonlinear, yielding spurious harmonics which must be rejected, along with outputs at the carrier frequency, this is done by low-pass filtering, which may also include video amplification if necessary. In contrast to video or baseband signals are carrier or bandpass signals whose spectrum is centered at some carrier frequency and does not include the frequency origin.

[†] There was sometimes also an audio signal in which the receiver output was converted to audible tones conveyed to the operator by a head set or a loudspeaker.

In laser radar, the detector is a photosensitive surface which produces an electrical current in response to the incident optical signal. In such detectors there is an option not ordinarily available in radar or sonar. Because of the extremely short optical wavelength, the angular resolution is such that individual scattering centers on the scattering object can be resolved, and an image of the object can therefore be formed. This is accomplished by employing a mosaic of discrete independent sensor elements—which are usually charge-coupled devices (CCDs)—in the photodetection plane, the dimensions of each such element being matched to the angular resolution width of the system. These elements are termed pixels.

The text is organized as follows. Chapters 2 and 3 present a review of probability, and noise and random processes. Although this material is self-contained, it is not meant to be a substitute for a first course in these subjects which, in addition to some exposure to Fourier transforms, would be advisable. Chapter 4 deals with conversion of continuous-time to discrete-time signals, the sampling theorem, the need for oversampling, and the special problems that arise in the processing of discrete-time signals when Fast Fourier Transform (FFT) filtering is employed. The relationship between the duration of a signal and its bandwidth is also discussed in this chapter and the analytic signal formulation is introduced. Detection of signals in the presence of noise is dealt with in Chapter 5, within the context of statistical decision theory as applied to a binary digital channel employing the likelihood ratio test, for which the Bayes, maximum likelihood and Neyman–Pearson decision criteria are each discussed. Matched filtering is introduced here. Both the Gaussian, and Poisson shot-noise channels are treated.

A comparison of coherent and noncoherent detection and processing is presented in Chapter 6, in which the important difference between the coherent matched filter and filtering with matching in amplitude only, and the associated difference in signal-to-noise ratio, is discussed. The necessary analytic expressions are derived and sufficient curves and tables are presented to enable selection of system parameters to achieve a specified level of performance in terms of detection and false alarm probabilities (or bit error rates). In particular, the question of predetection vs. postdetection integration is discussed in detail. Parameter estimation is dealt with in Chapter 7 within the context of generalized statistical estimation theory and the limits prescribed by the Cramer–Rao lower bound on the variance of an estimator. The method of maximum likelihood is introduced and applied to estimation of the position in range and angle, and of velocity, of a target in the presence of Gaussian noise. The results are also applied to target tracking and estimation of error in predicting target position. Chapter 8 deals with waveform analysis in terms of range and velocity resolution capability and with the measurement ambiguity inherent in periodically pulsed waveforms. The ambiguity diagram and its properties are discussed, including a generalized form of the diagram which may be required in sonar applications. The distortion of the transmitted signal that occurs with scattering from a moving target is discussed, together with the differences in these effects for radar and sonar systems because of differences in v/c. Chapter 9 continues this discussion, and introduces the subject of large time-bandwidth-product waveforms, including linear FM (chirp), hyperbolic FM, and pseudorandom signals. Chapter 10 presents a generalized discussion of coherent detection and processing, and Chapter 11 deals with basic system issues which arise in the design and operation of sensing systems. These include beampatterns and gain of antennas and arrays, the radar and sonar equations, and issues that arise during search, including the effect of false alarm probability and criteria for its specification.

2

REVIEW OF PROBABILITY

2.1 BERNOULLI TRIALS—THE BINOMIAL DISTRIBUTION

An experiment with two possible outcomes is performed in which a certain event either does occur (a success) or does not occur (a failure). The probability of success or failure on any given trial is respectively p and $q = 1 - p$. If n independent trials are performed there are

$$\binom{n}{k} = \frac{n!}{k!(n-k)!} \tag{2.1}$$

possible ways for k successes to occur, and the probability of k successes in n trials $P(k, n)$ is

$$P(k, n) = \binom{n}{k} p^k (1-p)^{n-k} \tag{2.2}$$

The binomial theorem

$$(a+b)^n = \sum_{k=0}^{n} \binom{n}{k} a^n b^{n-k} \tag{2.3}$$

yields the proper normalization

$$(p+q)^n = 1 = \sum_{k=0}^{n} \binom{n}{k} p^k (1-p)^{n-k} \tag{2.4}$$

and (2.2) is called the binomial distribution.

In dealing with the subject of probability we speak of a random variable R which is a function whose value depends on the outcome of an experiment. The statistical properties of R are described by its probability distribution. For a discrete random variable R with probability distribution $P_R(k)$ the expected value of an arbitrary function $g(R)$ of R, is†

$$E[g(R)] = \sum_k g(k) P_R(k) \tag{2.5}$$

and for a continuous random variable

$$E(g(R)] = \int_{-\infty}^{\infty} g(x) P_R(x)\, dx \tag{2.6}$$

with $P_R(k) \geq 0$, $P_R(x) \geq 0$ and of course

$$\sum_k P_R(k) = 1$$

$$\int_{-\infty}^{\infty} P_R(x)\, dx = 1$$

For Bernoulli trials, let the random variable r_i equal the number of successes on the ith trial. Then $r_i = 1$ with probability p and $r_i = 0$ with probability q. Hence, by (2.5), the expected values of r_i and r_i^2 are

$$E(r_i) = 1 \cdot p + 0 \cdot q = p \tag{2.7}$$
$$E(r_i^2) = 1^2 \cdot p + 0^2 \cdot q = p$$

The number of successes S in n trials is a random variable

$$S = \sum_{i=1}^{n} r_i \tag{2.8}$$

And the mean λ and variance $\text{Var}(S)$ of the Binomial distribution are therefore

$$\lambda = E(S) = E\left(\sum_{i=1}^{n} r_i\right) = \sum_{i=1}^{n} E(r_i) = np \tag{2.9}$$
$$\text{Var}(S) = E(S - E(s))^2 = n \times [E(r_i^2) - (E(r_i))^2] = nqp$$

where we recall that the variance of a random variable can also be expressed as the expected value of its square minus the square of its expected value.

The standard deviation of a random variable is equal to the square root of the variance. If an experiment is performed a large number of times we generally expect that the results will be reasonably centered around some

† Letting $g(R) = 1$ for $a \leq R \leq b$ and zero otherwise yields the probability that $a \leq R \leq b$.

average value, as determined by the governing probability distribution. There will however always be some spread in the values of the outcomes of the experiment around the average value, of which the standard deviation is a measure of what this spread might be expected to be. A measure of the significance of this spread is given by the ratio of the standard deviation to the mean. Clearly we would expect a standard deviation of, say, unity to be quite significant in an experiment in which the average value was 2, and much less important if the average value were 1000. For the binomial distribution the ratio of the standard deviation to the mean is $\sqrt{q/np}$. Thus, as the number of trials increases, the mean value np becomes a better estimate of the outcome, because the relative magnitude of the fluctuations around the mean become smaller. In this case the values of p and q of course also play a part.

2.2 THE POISSON DISTRIBUTION

Given the average number of successes λ in an experiment consisting of n Bernoulli trials, with $p = \lambda/n$, consider the limit of the Binomial distribution as n becomes very large and p becomes very small.

$$\lim_{n\to\infty} P(k, n) = \lim_{n\to\infty} \frac{n!}{k!(n-k)!} \left(\frac{\lambda}{n}\right)^k \left(1 - \frac{\lambda}{n}\right)^{n-k}$$

But

$$\lim_{n\to\infty} \frac{n!}{k!(n-k)!} \left(\frac{\lambda}{n}\right)^k = \lim_{n\to\infty} \frac{n(n-1)\cdots(n-k+1)\lambda^k}{k!n^k} = \frac{\lambda^k}{k!}$$

$$\lim_{n\to\infty} \left(1 - \frac{\lambda}{n}\right)^{-k} \left(1 - \frac{\lambda}{n}\right)^n = e^{-\lambda}$$

which yields the Poisson distribution

$$\lim_{n\to\infty} P(k, n) = \frac{\lambda^k}{k!} e^{-\lambda} \tag{2.10}$$

which gives the probability of k successes in an experiment in which the average number of successes is λ (the Poisson parameter). If the experiment takes place over a time T with λ the average number of times *per second* that the event occurs, then

$$P(k \text{ successes in } T \text{ s}) = \frac{(\lambda T)^k e^{-\lambda T}}{k!} \tag{2.11}$$

For the Poisson distribution the variance is equal to the mean. That is, $E(k) = E[k - E(k)]^2$ (see exercise 2.2), and the distribution is therefore completely specified by the single parameter λ. Thus, in the foregoing case the ratio of the standard deviation to the mean, the significance of which has been discussed in Section 2.1, is $(\lambda T)^{1/2}/\lambda T = 1/(\lambda T)^{1/2}$.

2.3 THE EXPONENTIAL DISTRIBUTION

Let λ be the probability per second that an event takes place; that is, the probability that the event takes place in some interval Δt is $\lambda\,\Delta t$. Suppose the event just occurs. To determine how much time will elapse before it occurs again, write the waiting time t as $t = n\,\Delta t$. Then, assuming independent events, the probability that t seconds will elapse before the event occurs is $(1-\lambda\,\Delta t)^n \lambda\,\Delta t$, and

$$\lim_{n\to\infty} (1-\lambda\,\Delta t)^n \lambda\,\Delta t = \lim_{n\to\infty} \left(1 - \frac{\lambda t}{n}\right)^n \lambda\,\Delta t = \lambda e^{-\lambda t}\,dt \tag{2.12}$$

Thus the probability that the second event will take place immediately after the first, $P(t=0)$, is $\lambda\,dt$ as expected, and the probability of longer waiting times decreases exponentially. The average waiting time is

$$\lambda \int_0^\infty t e^{-\lambda t}\,dt = \frac{1}{\lambda} \tag{2.13}$$

2.4 THE GAUSSIAN DISTRIBUTION

The probability distribution of a Gaussian random variable with mean μ and variance σ^2 is

$$P_R(x) = \frac{1}{\sqrt{2\pi}\sigma}\, e^{-(x-\mu)^2/2\sigma^2} \tag{2.14}$$

and $P\{a \le R \le b\}$, the probability that $a \le R \le b$ is

$$P\{a \le R \le b\} = \frac{1}{\sigma\sqrt{2\pi}} \int_a^b e^{-(x-\mu^2)/2\sigma^2}\,dx \tag{2.15}$$

This distribution is of particular importance because of the central limit theorem, which is proved below using characteristic functions. In speaking of R we say that R is Gaussian, or normal, (μ, σ).

2.5 THE RAYLEIGH AND RICE DISTRIBUTIONS

The Gaussian distribution is very important in radar and sonar because it describes the statistics of thermal noise and ambient ocean noise, which is therefore referred to as Gaussian noise. The Rayleigh and Rice distributions describe the statistics of the interference that is produced when the nonlinear operations frequently implemented in receiver systems are applied to Gaussian noise. These distributions however are more appropriately discussed in Chapter 3.

2.6 JOINT DISTRIBUTIONS, CONDITIONAL DISTRIBUTIONS, AND BAYES THEOREM

For two random variables R_1 and R_2, the probability that $a \leq R_1 \leq b$ and $c \leq R_2 \leq d$ is

$$P\{a \leq R_1 \leq b \text{ and } c \leq R_2 \leq d\} = \int_a^b dx \int_c^d dy \, P_{R_1R_2}(x, y) \tag{2.16}$$

where $P_{R_1R_2}(x, y)$ is the joint distribution of R_1 and R_2, and

$$P_{R_1}(x) = \int_{-\infty}^{\infty} P_{R_1R_2}(x, y) \, dy \tag{2.17}$$

If R_1 and R_2 are independent

$$P_{R_1R_2}(x, y) = P_{R_1}(x) P_{R_2}(y) \tag{2.18}$$

Dropping the subscripts R_1 and R_2 for convenience, the conditional distribution of x given y is

$$P(x|y) = \frac{P(x, y)}{P(y)} \tag{2.19}$$

which of course is defined only if $P(y) \neq 0$. If x and y are independent, then from (2.18) $P(x|y) = P(x)$, as expected.

From (2.19), interchanging x and y yields Bayes' Theorem

$$P(y|x) = \frac{P(y)P(x|y)}{P(x)} \tag{2.20}$$

which is very important in signal-detection theory. Suppose x represents the statistical information available to a detection system as the result of observations, and y represents the statistical condition regarding the presence or absence of a target of interest. Equation (2.20) gives statistical information regarding the likelihood of a target being present or absent, given the observations (a posteriori conditional distribution $P(y|x)$), in terms of the a priori conditional distribution relating what is likely to be observed when the target is or is not present, $P(x|y)$, which can be calculated prior to the observations. Equation (2.20) is the basis for Chapter 5.

With the use of the conditional distribution the expected value of x given y, $E(x|y)$, for the discrete and continuous cases is defined as

$$E(x|y) = \sum_x xP(x|y) \tag{2.21}$$

and

$$E(x|y) = \int xP(x|y)\,dx$$

from which $E(x)$ can be expressed as a weighted average of conditional expectation. That is, from (2.19) and (2.21)

$$E(x) = \sum_y E(x|y)P(y) \tag{2.22}$$

or

$$E(x) = \int E(x|y)P(y)\,dy$$

$P(x)$ can also be expressed as a weighted average of conditional probabilities:

$$P(x) = \sum_y P(y)P(x|y) \tag{2.23}$$

which is sometimes convenient.

2.7 CHARACTERISTIC FUNCTIONS

The characteristic function $M_R(f)$ of a random variable R is

$$M_R(f) = E(e^{-i2\pi Rf}) = \int_{-\infty}^{\infty} P_R(x)e^{-i2\pi fx}\,dx \tag{2.24}$$

the Fourier transform of $P_R(x)$.

The characteristic function is useful for generating the moments of $P_R(x)$. That is, the first moment of $P_R(x)$, $E(x) = \mu$, is

$$\mu_1 = E(x) = \left(\frac{i}{2\pi}\right) \frac{dM_R(f)}{dx}\bigg|_{f=0} = \int_{-\infty}^{\infty} xP_R(x)\,dx \tag{2.25}$$

And in general

$$\left(\frac{i}{2\pi}\right)^n \frac{d^n M_R(f)}{df^n}\bigg|_{f=0} = E(x^n) = \mu_n \tag{2.26}$$

We shall often write the first moment as μ, leaving out the subscript 1. In the terms of moments the variance of R is $\mathrm{Var}(R) = \mu_2 - \mu^2$.

From (2.18), the joint distribution of n independent random variables (RVs) is the product of the n distributions:

$$P_{R_1 R_2 \cdots R_n}(x_1, x_y \cdots x_n) = P_{R_1}(x_1) P_{R_2}(x_2) \cdots P_{R_n}(x_n) \tag{2.27}$$

Suppose $R_3 = R_1 + R_2$, where R_1 and R_2 are independent. Then

$$M_{R_3}(f) = E(e^{-i2fR_3}) = E(e^{-i2\pi f R_1}) E(e^{-i2\pi f R_1}) = M_{R_1}(f) M_{R_2}(f) \tag{2.28}$$

by (2.27), and the distribution of R_3 is

$$\begin{aligned} P_{R_3}(z) &= \int_{-\infty}^{\infty} M_{R_3}(f) e^{i2\pi fz}\, df \\ &= \int_{-\infty}^{\infty} M_{R_1}(f) M_{R_2}(f) e^{i2\pi fz}\, df = \int_{-\infty}^{\infty} P_{R_1}(x) P_{R_2}(z - x)\, dx \end{aligned} \tag{2.29}$$

Thus, the probability density function of the sum of n independent RVs is given by n convolutions of the respective densities.

The characteristic function of a Gaussian random variable is

$$M_R(f) = \int_{-\infty}^{\infty} \exp\left[-\frac{(x - \mu)^2}{2\sigma^2} \right] \exp(-i2\pi fx) \frac{dx}{\sqrt{2\pi}\sigma} \tag{2.30}$$

By completing the square in the exponent

$$\begin{aligned} &-\frac{1}{2\sigma^2} [(x - \mu)^2 + i2\pi 2\sigma^2 fx] \\ &= -\frac{1}{2\sigma^2} [(x - \mu + i2\pi\sigma^2 f)^2 + i2\pi\sigma^2 2\mu f + 4\pi^2\sigma^4 f^2] \end{aligned} \tag{2.31}$$

And

$$M_R(f) = \exp(-i2\pi\mu f - 2\pi^2\sigma^2 f^2) \int_{-\infty}^{\infty} \exp\left[-\frac{(x - \mu + i2\pi\sigma^2 f)^2}{2\sigma^2} \right] \frac{dx}{\sqrt{2\pi}\sigma} \tag{2.32}$$

Since there are no poles enclosed within the contour in Figure 2.1, then

$$\int_{-\infty}^{\infty} e^{-(x-\mu-i2\pi\sigma^2 f)^2/2\sigma^2} \frac{dx}{\sigma\sqrt{2\pi}} = \int_{-\infty}^{\infty} e^{-(x-\mu)^2/2\sigma^2} \frac{dx}{\sigma\sqrt{2\pi}} = 1$$

Hence

$$M_R(f) = e^{-i2\pi\mu f - 2\pi^2\sigma^2 f^2} \tag{2.33}$$

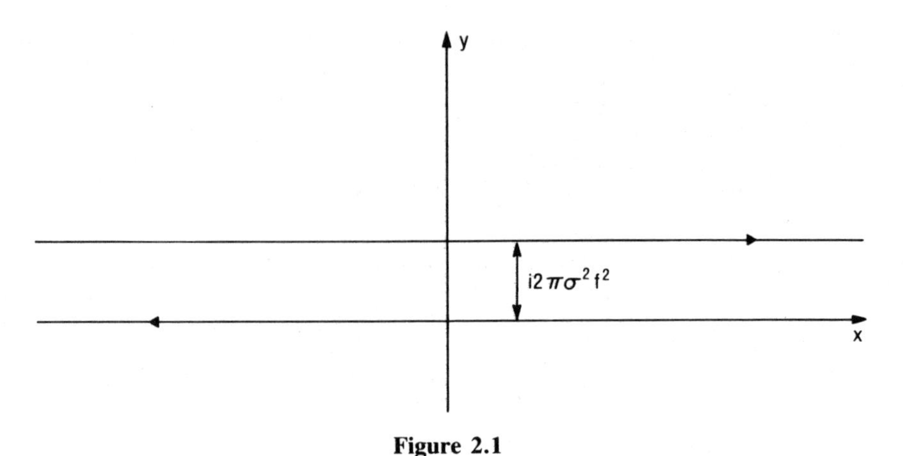

Figure 2.1

From this we observe that the sum of any number of independent Gaussian RVs remains Gaussian with mean equal to the sum of the means, and variance equal to the sum of the variances. Also, the convolution of any number of Gaussian functions is Gaussian, with the same condition holding on the mean and variance of the result of the convolution. In general, the result of any linear operation on Gaussian random variables is Gaussian.

2.8 THE LAW OF LARGE NUMBERS

Let R_i denote the value taken by the random variable R on the ith trial of an experiment, with for all i

$$E(R) = E(R_i) = \mu \tag{2.34}$$

$$\text{Var}(R) = E(R_i^2) - \mu^2 = \sigma^2$$

Define the random variable S_n as

$$S_n = \frac{1}{n} \sum_{i=1}^{n} R_i \tag{2.35}$$

then

$$E(S_n) = \frac{1}{n} E\left(\sum_{i=1}^{n} R_i\right) = \frac{n\mu}{n} = \mu \tag{2.36}$$

that is the law of averages, and it is left as an exercise to show that

$$\text{Var}(S_n) = \sigma_n^2 = \frac{1}{n^2} E\left(\sum_{i=1}^{n} R_i\right)^2 - \mu^2 = \frac{\sigma^2}{n} \tag{2.37}$$

Thus, as n becomes very large the random variable S_n—which also can be thought of as a limit of a sequence of random variables $S_1, S_2, \ldots, S_n$—approaches a *constant* μ; that is, as $n \to \infty$ the variance of σ_n^2 of S_n approaches zero. This is the basis for many practical signal-processing schemes in active sensing systems and many other applications.

As a consequence of (2.36) and (2.37) it can be shown that for any $\epsilon > 0$

$$\lim_{n \to \infty} P\{|S_n - \mu|\} > \epsilon = 0 \tag{2.38}$$

which is the formal statement of the weak law of large numbers.

2.9 THE CENTRAL LIMIT THEOREM

Let $R_1, R_2, \ldots, R_n$ be n independent identically distributed random variables with first two moments finite, i.e. for all i

$$E(R_i) = \int_{-\infty}^{\infty} x P_{R_i}(x)\, dx = \mu_1 < \infty \tag{2.39}$$

$$E(R_i^2) = \int_{-\infty}^{\infty} x^2 P_{R_i}(x)\, dx = \mu_2 < \infty \tag{2.40}$$

Define

$$S_n = \sum_{i=1}^{n} R_i \tag{2.41}$$

in which case

$$\begin{aligned} E(S_n) &= n\mu \\ \mathrm{Var}(S_n) &= n\,\mathrm{Var}(R) = n\sigma^2 \end{aligned} \tag{2.42}$$

Let

$$X_n = \frac{S_n - n\mu}{\sqrt{n}\sigma} \tag{2.43}$$

in which

$$\begin{aligned} E(X_n) &= 0 \\ \mathrm{Var}(X_n) &= \frac{n^2\sigma^2}{n^2\sigma^2} = 1 \end{aligned} \tag{2.44}$$

The central limit theorem states that as $n \to \infty$ the probability density of X_n approaches the Gaussian (normal) distribution, with zero mean and unit variance, that is

$$\lim_{n\to\infty} P_{X_n}(x) \to \frac{1}{\sqrt{2\pi}} e^{-x^2/2} \tag{2.45}$$

This is easily proved using characteristic functions. For the random variable X_n

$$\begin{aligned}
M_{X_n}(f) &= E\left\{\exp\left[-i2\pi f\left(\frac{S_n - n\mu}{\sqrt{n}\sigma}\right)\right]\right\} \\
&= \exp\left(i2\pi \frac{\sqrt{n}\mu f}{\sigma}\right) E\left[\exp\left(-i2\pi f \frac{\Sigma_j R_j}{\sqrt{n}\sigma}\right)\right] \\
&= \exp\left(i2\pi \frac{\sqrt{n}\mu f}{\sigma}\right)\left\{E\left[\exp\left(-i \frac{2\pi f R}{\sqrt{n}\sigma}\right]\right\}^n \\
&= \exp\left(i2\pi \frac{\sqrt{n}\mu f}{\sigma}\right)\left[M_R\left(\frac{f}{\sqrt{n}\sigma}\right)\right]^n
\end{aligned} \tag{2.46}$$

where the last two steps follow from (2.27) and (2.28). Then

$$\log M_{X_n}(f) = i2\pi \frac{\sqrt{n}\mu}{\sigma} f + n \log M_R\left(\frac{f}{\sqrt{n}\sigma}\right) \tag{2.47}$$

Now, by expanding $M_R(f/\sqrt{n}\sigma)$ in a Taylor series about $f = 0$ and using (2.26)

$$M_R\left(\frac{f}{\sqrt{n}\sigma}\right) = \left[1 - \frac{i2\pi\mu f}{\sqrt{n}\sigma} - \frac{4\pi^2\mu_2}{n\sigma^2}\frac{f^2}{2} + \mathcal{O}\left(\frac{1}{n^{3/2}}\right)\right] \tag{2.48}$$

and we can write, by making use of the expansion:

$$\log(1 - x) = -x - \frac{x^2}{2} - \cdots \tag{2.49}$$

$$n \log M_R\left(\frac{f}{\sqrt{n}\sigma}\right) \approx -i2\pi \frac{\sqrt{n}\mu f}{\sigma} - \frac{4\pi^2\mu_2}{\sigma^2}\frac{f^2}{2} + \frac{4\pi^2\mu^2}{2\sigma^2} f^2 + \mathcal{O}\left(\frac{1}{n^{1/2}}\right) \tag{2.50}$$

Thus,

$$\log M_{X_n}(f) = -2\pi^2 f^2 \frac{\mu_2 - \mu^2}{\sigma^2} = -2\pi^2 f^2 + \mathcal{O}\left(\frac{1}{n^{1/2}}\right) \tag{2.51}$$

since $\mathrm{Var}(R) = \sigma^2 = \mu_2 - \mu^2$ and

$$\lim_{n\to\infty} M_{X_n}(f) = e^{-2\pi^2 f^2} \tag{2.52}$$

which may be compared with (2.33) with $\mu = 0$ and $\sigma^2 = 1$.

2.10 APPROXIMATIONS TO THE GAUSSIAN DISTRIBUTION

Both the Poisson and Binomial distributions can be approximated by the Gaussian distribution under appropriate conditions. For the Poisson distribution

$$M_R(f) = E(e^{-i2\pi Rf}) = \sum_{k=0}^{\infty} e^{-\lambda} \frac{\lambda^k e^{-i2\pi fk}}{k!}$$

$$= e^{-\lambda} \sum_{k=0}^{\infty} \frac{(\lambda e^{-i2\pi f})^k}{k!} = \exp[\lambda(e^{-i2\pi f} - 1)] \tag{2.53}$$

The exponent is

$$\lambda(e^{-i2\pi f} - 1) = -i\lambda \sin 2\pi f - \lambda(1 - \cos 2\pi f) = -i\lambda \sin 2\pi f - 2\lambda \sin^2 \pi f \tag{2.54}$$

and

$$M_R(f) = e^{-i\lambda \sin 2\pi f} e^{-2\lambda \sin^2 \pi f} \tag{2.55}$$

Now if λ is sufficiently large then $M_R(f)$ is negligible for all but very small values of $\sin \pi f$, in which case $\sin 2\pi f \sim 2\pi f$ and $\sin^2 \pi f \sim \pi^2 f^2$ and

$$M_R(f) \approx e^{-i2\pi\lambda f} e^{-2\pi^2 \lambda f^2} \tag{2.56}$$

which may be compared with (2.33) with $\lambda = \mu$ and $\lambda = \sigma^2$. Hence,

$$\frac{e^{-\lambda}\lambda^k}{k!} \sim \frac{1}{\sqrt{2\pi\lambda}} \exp\left[-\frac{(k-\lambda)^2}{2\lambda}\right] \tag{2.57}$$

Note that the Poisson distribution in the approximation by a Gaussian remains a single-parameter distribution.

For the Binomial distribution

$$M_R(f) = E(e^{-i2\pi fR}) = \sum_{k=0}^{n} \binom{n}{k} e^{-i2\pi fk} p^k (1-p)^{n-k} = (pe^{-i2\pi f} + q)^n \tag{2.58}$$

from the Binomial theorem (2.3).

Now the term inside the parentheses on the right-hand side of (2.58) is a complex number $re^{-i\theta}$ as shown in Figure 2.2 for which, since $\sigma^2 = nqp$

$$r^2 = p^2 + q^2 + 2pq \cos 2\pi f = p^2 + q^2 + 2pq(1 - 2\sin^2 \pi f)$$

$$= 1 - 4pq \sin^2 \pi f = 1 - \frac{4\sigma^2}{n} \sin^2 \pi f \tag{2.59}$$

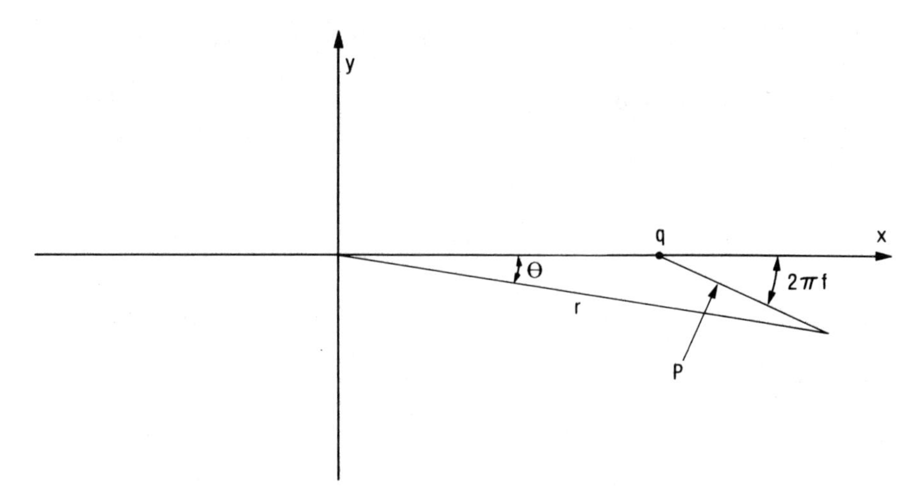

Figure 2.2

Hence, in the limit as n becomes large

$$r^n = \left(1 - \frac{4\sigma^2 \sin^2 \pi f}{n}\right)^{n/2} \sim e^{-2\sigma^2 \sin^2 \pi f} \tag{2.60}$$

Once again we make the argument that if npq is large then $M_R(f)$ will be nonnegligible only for very small values of $\sin^2 \pi f \sim \pi^2 f^2$ and

$$r^n \sim e^{-2\pi^2 \sigma^2 f^2} \tag{2.61}$$

For the phase

$$\theta = -\tan^{-1} \frac{p \sin 2\pi f}{q + p \cos 2\pi f} \sim -2\pi f p \tag{2.62}$$

for $\pi f \ll 1$. Hence, for large npq, and using $\mu = np$

$$M_R(f) \approx r^n e^{in\theta} \approx e^{-i2\pi\mu f} e^{-2\pi^2 \sigma^2 f^2} \tag{2.63}$$

and from (2.9)

$$P(k, n) \approx \frac{\exp[-(k - np)^2/2npq]}{\sqrt{2\pi npq}} \tag{2.64}$$

A summary of the various distributions which have been discussed, and conditions under which they can be approximated by the Poisson and Gaussian distributions, is given in Table 1.

TABLE 1

Distribution	Conditions	Approximation
Binomial $P(k, n) = \binom{n}{k} p^k q^{n-k}$	n large p small	Poisson with $\lambda = np$
Binomial	$np(1-p)$ large	Gaussian with $\mu = np$, $\sigma^2 = npq$
Poisson	λ or λT large	Gaussian with $\mu = \sigma^2 = \lambda$ or λT

The question of how large or how small these parameters must be depends on the application. Certainly, the approximations could be expected to be valid for, say, $n \geq 100$, $p \leq 0.01$, $np(1-p) > 100$, and λ or $\lambda T \geq 100$. But, often, less stringent conditions may be satisfactory. For example, in laser radar a value of $\lambda T \sim 5$ photons is often sufficient to justify the Gaussian approximation for the Poisson distribution.

2.11 FUNCTIONS OF A RANDOM VARIABLE

Consider a probability density function (pdf) $P_R(x)$. Then $P\{x \leq R \leq x + \Delta x\} = P_R(x)\, dx$. Now suppose $y = f(x)$. Then clearly $P_R(x)\, dx = P_S(y)\, dy$ where $S = f(R)$. That is, with the change of variables $y = f(x)$ we maintain $P\{x \leq R \leq x + \Delta x\} = P\{y \leq S \leq y + \Delta y\}$. The transformation of the pdf with the change in variables is therefore

$$P_S(y) = \frac{P_R(x)}{\left|\dfrac{dy}{dx}\right|} \tag{2.65}$$

This transformation can be extended to pdfs of arbitrary dimension as follows. If $P_R(x_1, x_2, \ldots, x_n)$ is an n-dimensional pdf and if $y_1 = y_1(x_1, x_2, \ldots, x_n)$, $y_2 = y_2(x_1, x_2, \ldots, x_n), \ldots$. Then as before

$$\begin{aligned} &P_R(x_1, x_2, \ldots, x_n)\, dx_1, dx_2, \ldots, dx_n \\ &\quad = P_S(y_1, y_2, \ldots, y_n)\, dy_1, dy_2, \ldots, dy_n \end{aligned} \tag{2.66}$$

But with a change of variables, the pdfs therefore transform according to

$$P_S(y_1, y_2, \ldots, y_n) = \frac{P_R(x_1, x_2, \ldots, x_n)}{J\left(\dfrac{\partial y}{\partial x}\right)} \tag{2.67}$$

where $J(\partial y/\partial x)$ is the Jacobian of the transformation:

$$J\left(\frac{\partial y}{\partial x}\right) = \det \begin{pmatrix} \frac{\partial y_1}{\partial x_1} & \frac{\partial y_2}{\partial x_1} & \cdots & \frac{\partial y_n}{\partial x_1} \\ \vdots & & & \vdots \\ \frac{\partial y_1}{\partial x_n} & \frac{\partial y_2}{\partial x_n} & \cdots & \frac{\partial y_n}{\partial x_n} \end{pmatrix}$$

EXERCISES FOR CHAPTER 2

2.1 Referring to (2.8) show by direct calculation (using (2.2)) that $E(S) = np$ and $\mathrm{Var}(S) = nqp$.

2.2 Show for the Poisson distribution (2.10) that $E(k) = \mathrm{Var}(k) = \lambda$.

2.3 Show using (2.19) and (2.21) that (2.22) reduces to (2.5).

2.4 Prove that the sum of any number of independent Gaussian random variables is Gaussian, with mean equal to the sum of the means, and variance equal to the sum of the variances.

2.5 Prove that the convolution of any number of Gaussian functions is Gaussian.

2.6 Show that the Bayes theorem can be written

$$P(x|y) = \frac{P(x)P(y|x)}{\int P(x)P(y|x)\, dx}$$

or

$$P(x|y) = \frac{P(x)P(y|x)}{\Sigma_x P(x)P(y|x)}$$

2.7 Prove (2.37), that $\mathrm{Var}(S_n) = \sigma^2/n$.

2.8 Prove (2.44), that $E(X_n) = 0$ and $\mathrm{Var}(X_n) = 1$.

2.9 Find the mean and variance of the exponential distribution $\lambda e^{-\lambda t}$ by direct calculation and by use of the characteristic function.

2.10 An event has an average rate of occurrence λ. Write an expression for the probability that at the event occurs at least k_0 times in T seconds. If $\lambda = 10^{-3}$ and $T = 90$ s calculate the probability that the event occurs at least three times during this interval.

2.11 Show that

$$\int P_R(x|z)\, dx = 1$$

for the continuous random variable R.

2.12 Show that

$$P(u, v|w) = P(u|v, w)P(v|w)$$

2.13 If $R = \log S$ is Gaussian (μ, σ), then S is said to be log normal. Show that

$$P_S(y) = \frac{1}{\sqrt{2\pi}\sigma y} e^{-(\ln y - \mu)^2/2\sigma^2} y \geq 0$$

and that the first and second moments are

$$E(S) = e^{\mu + \frac{1}{2}\sigma^2}$$

$$E(S^2) = e^{2\mu + 2\sigma^2}$$

2.14 Prove the generalization of Bayes theorem

$$P(x_i|y_j) = \frac{P(x_i)P(y_j|x_i)}{\Sigma_k P(x_k)P(y_j|x_k)}$$

3

REVIEW OF NOISE AND RANDOM PROCESSES

3.1 INTRODUCTION—CORRELATION FUNCTIONS AND POWER SPECTRAL DENSITIES

The previous chapter dealt with random variables, which are functions that assume specific values with certain probabilities depending on the outcome of an experiment. We now deal with random time functions. A random process can be thought of as a system which in the course of an experiment produces one of an ensemble of time functions. With a random process the outcome of the experiment is a function of time. A random process can therefore be represented as an ensemble of functions $x(t, \alpha)$, with a different time function corresponding to each value of the parameter α which is selected randomly by the process. A particular outcome $x(t, \alpha)$ is referred to as a realization of the random process. For a continuous variable we have a continuous random process; for a discrete random process the functions are $x(k, \alpha)$ where k takes discrete values. As an example of a continuous random process consider the family of time functions $\cos(2\pi ft + \theta)$ with θ uniformly distributed over $(0, 2\pi)$. In this example the parameter α is represented by θ, and once the experiment is performed the outcome is a time function $\cos(2\pi ft + \theta)$. On the other hand, the process can also consist of an ensemble of time functions that cannot be so described analytically, in which any given realization itself appears to be random in time; for example, Gaussian noise. As far as the general definition of a random process is concerned, such distinctions are of course immaterial.

For a random process, in the most general case the mean is of the form

$$E(x(t, \alpha)) = \mu(t) = \int P_t(\alpha)x(t, \alpha)\, d\alpha \tag{3.1}$$

That is, for a given value of t the average is taken over all values of $x(t, \alpha)$ in the ensemble, and the result can be different for different values of t. Also, the probability distribution $P_t(\alpha)$ can vary with t. In what follows however we shall not be concerned with the most general case, but with a subset, stationary random processes. The necessary and sufficient conditions for a process to be (wide-sense)[†] stationary are

1. Time-invariance of the mean:

$$\int x(t, \alpha) P(\alpha)\, d\alpha = \mu \tag{3.2}$$

   where μ is constant and $P(\alpha)$ is also independent of time.
2. The autocorrelation function[‡] $E[x(t_1, \alpha)x(t_2, \alpha)]$ must be of the form

$$E[x(t_1, \alpha)x(t_2, \alpha)] = \int x(t_1, \alpha)x(t_2, \alpha)P_{12}(\alpha)\, d\alpha = r(t_2 - t_1) \tag{3.3}$$

The autocorrelation function is a measure of the degree of dependence between the random variables $x(t_1, \alpha)$ and $x(t_2, \alpha)$ in the sense that if $E[x(t_1, \alpha)] = E[x(t_2, \alpha)] = 0$ and $x(t_1, \alpha)$ and $x(t_2, \alpha)$ are independent, then $r(t_2 - t_1) = E[x(t_1, \alpha)x(t_2, \alpha)] = E[x(t_1, \alpha)]E[x(t_2, \alpha)] = 0$. Equation (3.3) states that for a stationary process the degree of dependence between the random variables $x(t_1, \alpha)$ and $x(t_2, \alpha)$ depends only on the interval $t_2 - t_1$, independent of the choice of time origin. The joint probability distribution $P_{12}(\alpha)$ of the random variables $x(t_1, \alpha)$ and $x(t_2, \alpha)$, again, must be independent of the specific values of t_2 and t_1, but can be a function of $t_2 - t_1$.

Writing $t_1 = t$, $t_2 = t + \tau$ one would expect intuitively that as τ gets very large the two random variables become independent and

$$\lim_{\tau\to\infty} E[x(t, \alpha)x(t + \tau, \alpha)] = E[x(t, \alpha)]E[x(t + \tau, \alpha)] = \mu^2 \tag{3.4}$$

This is often the case, but not always.

Henceforth we shall drop the notation α and write the realization as $x(t)$, with the understanding that $E[x(t)x(t + \tau)]$ denotes an average over the ensemble of time functions which make up the process. In addition to the correlation function obtained by an ensemble average, the time-autocorrelation function of a random function $x(t)$ can be defined as

$$\langle x(t)x(t + \tau)\rangle = \lim_{T\to\infty} \frac{1}{T} \int_{-T/2}^{T/2} x(t)x(t + \tau)\, dt \tag{3.5}$$

[†] There are also processes which are strictly stationary which we do not deal with here.
[‡] A cross-correlation function can also be defined, see Ex. (3.17).

In certain cases, ergodic processes, averaging over the ensemble is equivalent to averaging over time, and in this case, for an ergodic process:

$$E[x(t)x(t+\tau)] = \int_{-\infty}^{\infty} x(t, \alpha)x(t+\alpha, \alpha)P(\alpha)\, d\alpha = \langle x(t)x(t+\tau)\rangle \quad (3.6)$$

The power spectral density $W(f)$ of a random process is defined as

$$W(f) = \lim_{T\to\infty} E\left[\frac{X_T(f)X_T^*(f)}{T}\right] \quad (3.7)$$

where $X_T(f)$ is the truncated Fourier transform of the realization $x(t)$:

$$X_T(f) = \int_{-T/2}^{T/2} x(t)e^{-i2\pi ft}\, dt$$

and is therefore a random function of f.

By the Wiener Khinchine theorem (see Exercise 3.4)

$$W(f) = \int_{-\infty}^{\infty} r(\tau)e^{-i2\pi ft}\, d\tau \quad (3.8)$$

where $r(\tau)$ is the correlation function

$$r(\tau) = E[x(t)x(t+\tau)] \quad (3.9)$$

For discrete-time signals the autocorrelation function of a stationary process $x(n\,\Delta t)$ is $R(m, n) = E[x((m+n)\,\Delta t)x(n\,\Delta t)] = R(m)$, a function only of the time separation $m\,\Delta t$. The power spectral density $S(f)$ is defined as

$$S(f) = \lim_{N\to\infty} E\left[\frac{X_N(f)X_N^*(f)}{N}\right]\Delta t$$

where $X_N(f)$ is the truncated Fourier transform

$$X_N(f) = \sum_{n=-N/2}^{N/2} x(n\,\Delta t)e^{-i2\pi fn\,\Delta t}$$

And also

$$S(f) = \Delta t \sum_{m=-\infty}^{\infty} R(m)e^{-i2\pi mf\,\Delta t}$$

with

$$R(m) = \int_{-1/2\,\Delta t}^{1/2\,\Delta t} S(f)e^{i2\pi mf\,\Delta t}\, df$$

As a consequence of the Wiener Khinchine theorem, the average power $E[x^2(t)]$ can be expressed alternatively as

$$E[x^2(t)] = r(0) = \int_{-\infty}^{\infty} W(f)\, df \tag{3.10}$$

For a linear system with impulse response $h(t)$ and input $y(t)$ the output $x(t)$ is given by (see Section 4.5)

$$x(t) = \int_{-\infty}^{\infty} h(\tau)y(t-\tau)\, d\tau$$

from which it follows that the power spectral density $W_y(f)$ of the output $y(t)$ is given by (Exercise 3.5)

$$W_x(f) = |H(f)|^2 W_y(f) \tag{3.11}$$

where the transfer function $H(f)$ is

$$H(f) = \int_{-\infty}^{\infty} h(t)e^{-i2\pi ft}\, dt$$

and $W_y(f)$ is the power spectral density of the input. The average power of $x(t)$ is (see Exercise 3.5)

$$E[x^2(t)] = \int_{-\infty}^{\infty} W_x(f)\, df = \int_{-\infty}^{\infty} |H(f)|^2 W_y(f)\, df \tag{3.12}$$

An important relationship for Gaussian random variables is

$$\begin{aligned} E[x(t_1)x(t_2)x(t_3)x(t_4)] &= E[x(t_1)x(t_2)]E[x(t_3)x(t_4)] \\ &+ E[x(t_1)x(t_3)]E[x(t_2)x(t_4)] + E[x(t_1)x(t_4)]E[x(t_2)x(t_3)] \end{aligned} \tag{3.13}$$

For the special case $t_1 = t_2$, $t_3 = t_4 = t_1 + \tau$ this reduces to

$$E[x^2(t_1)x^2(t_1+\tau)] = 2r^2(\tau) + r^2(0) \tag{3.14}$$

3.2 TYPES OF NOISE

The most important types of noise in active sensing systems that can be treated statistically in a systematic way are thermal noise and shot noise. Thermal noise arises from random currents due to Brownian motion of electrons in receiver components such as resistors and, in fact, can arise only in lossy elements—we recall Kirchoff's comment on Black-Body radiation, that good emitters make good absorbers. Because the duration of the elementary current pulses which make up thermal noise correspond to the time between collisions of electrons, which is of the order of 10^{-14} s at room temperature, the power spectral density is flat over a very wide range of frequencies and said to be white, since, by analogy to light, all frequencies are equally represented. This is one of the major results of Nyquist's theorem, to be discussed, which also proves that thermal noise depends only on the temperature.

Radar is subject to a second kind of thermal noise, which might be termed ambient noise, that arises because of the finite temperature of the background viewed by the antenna. The antenna, approximately in thermal equilibrium with the background, produces thermal noise proportionately. It is for this reason that earth stations of communication-satellite systems, in which the antenna views for the most part a background of cold space, employ low-noise amplifiers in the front ends. On the other hand, in the satellites themselves, which are continually viewing the earth which is effectively a 300 K black body, there is no point in employing a front end cooled much below 300 K. In sonar there is also ambient noise which arises from random pressure waves that are always present. In both radar and sonar the statistical properties of the ambient noise are essentially the same as that of thermal noise and in the mathematical treatment that follows we make no distinction between ambient noise and thermal noise generated in the receiver.

Shot noise was first observed in vacuum tubes, in which the anode current was found to exhibit fluctuations due to variations in the emission times of electrons from the cathode. More recently, this effect has become very important in optical sensing systems since the rate of arrival of photons in a laser beam also exhibits random fluctuations about a mean value, and, as mentioned in Chapter 1, the rate of photon arrival in systems of practical interest can be sufficiently low such that these fluctuations are very much in evidence. Thus the signal itself incorporates randomness, which is fundamentally different from the thermal-noise situation in which a random quantity is added to a signal that may be purely deterministic. The term shot noise refers to the random fluctuations observed in the electronic currents produced by photodetectors as a result of the randomness in the arrival times of photons on the detector surface. These fluctuations are dependent only on the statistics of emission of electrons from the photodetector and are independent of temperature. Although optical systems are subject both to thermal as well as shot noise, the latter can be the dominant consideration, as will be seen in the discussion of optical heterodyne detection. Laser radar is also subject to a form of ambient noise which is more appropriately discussed in Chapter 5.

Other types of noise, which will not be considered further, are: *Flicker Noise*—$W(f) \propto 1/f$, which is a problem at low frequencies in semiconductors. $1/f$ noise is observed in many various phenomena; *Impulse Noise*—short, randomly occurring spikes that are often man-made; *Quantization*

Noise—This occurs when a continuous time waveform is approximated by a predetermined, finite, discrete number of voltage or current levels. The quantized signal can be viewed as the original continuous function plus quantization noise, which represents the uncertainty regarding the extent to which the quantized levels actually correspond to the original continuous values. Conversion of continuous to discrete-time signals is accomplished by means of analog-to-digital (A/D) converters which divide the full dynamic range of the signal of interest into some number m of quantized levels, usually expressed as $m = 2^n$. In this case, the signal-to-quantization-noise ratio is nominally $3 + 6n$ dB (e.g. 99 dB for $n = 16$), and quantization noise can therefore be eliminated from consideration if n is large enough.

3.3 POWER SPECTRAL DENSITY OF THERMAL NOISE—NYQUIST'S THEOREM

The power spectral density for thermal noise was derived by Nyquist in 1927 [10] by treating a conductor connecting two resistors in thermal equilibrium with their surroundings as a one-dimensional black body (Figure 3.1). Each resistor generates random currents, and in equilibrium the power generated by each resistor is equal to the power absorbed. We wish to find an expression for the power generated by each resistor as a function of frequency.

Recalling the approach used in black-body radiation, the fundamental mode of vibration in the conductor has wavelength $\lambda = 2L$ and frequency $f = v/2L$ where v is the velocity of propagation. Higher-order modes have frequencies $nv/2L$, $n = 2, 3, \ldots$; so for large L the number of modes Δn in a band Δf is $\Delta n = (2L/v)\,\Delta f$. Now from the Planck radiation law, the number of photons per mode of vibration is

$$\frac{1}{e^{hf/kT} - 1} \approx \frac{kT}{hf} \tag{3.15}$$

where h is Planck's constant (6.62×10^{-29} erg-s) and k is Boltzmann's

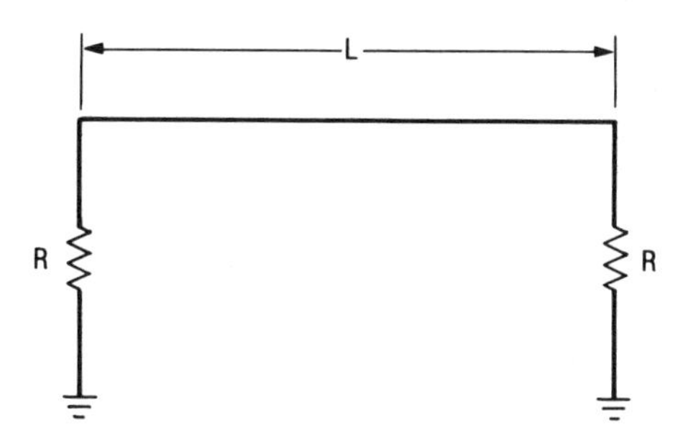

Figure 3.1

constant (1.38×10^{-16} erg/deg). The approximation in (3.15) holds for $f < 10^{13}$ Hz—below infrared frequencies. Since each photon has energy hf the total energy in Δf is $(2kTL/v)\,\Delta f$, the total energy per unit length is $(2kT/v)\,\Delta f$, and the amount of energy crossing any point per second in either direction—that is, the power produced by either resistor—is $kT\,\Delta f$. The combination of thermal-current-generator-resistor can be thought of as a voltage source V in series with R from which $V^2\,\Delta f = 4kTR\,\Delta f$.

Now consider the situation shown in Figure 3.2 where the resistor on the left is a constant at all frequencies and on the right there is a general frequency-dependent complex impedance. The power delivered to $R(f)$ by the equivalent voltage source in series with R is

$$\frac{V^2 R(f)\,\Delta f}{[R + R(f)]^2 + X^2(f)} \tag{3.16}$$

and for the equivalent voltage source $V(f)$ in series with $R(f)$ the power dissipated in R is

$$\frac{V^2(f) R\,\Delta f}{(R + R(f))^2 + X^2(f)} \tag{3.17}$$

But in equilibrium the relationship $V^2(f)R\,\Delta f = V^2 R(f)\,\Delta f$ must hold, and therefore

$$V^2(f)\,\Delta f = 4kTR(f)\,\Delta f \tag{3.18}$$

That is, the power that would be dissipated in a 1-ohm resistor in a band Δf resulting from random thermal currents generated in a frequency-dependent resistor $R(f)$ is $4kTR(f)\,\Delta f$; a function only of temperature and $R(f)$. This is known as (one of) Nyquist's theorem(s).

In most cases of practical interest there will be conjugate matching conditions as shown in Figure 3.3 where the left side might represent a radar antenna with equilibrium temperature T and the right side the input

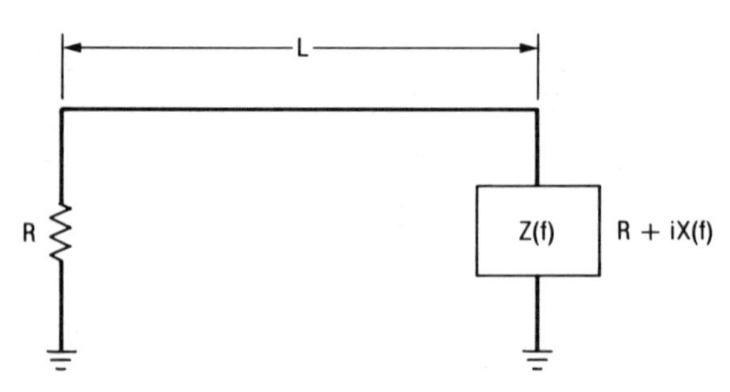

Figure 3.2

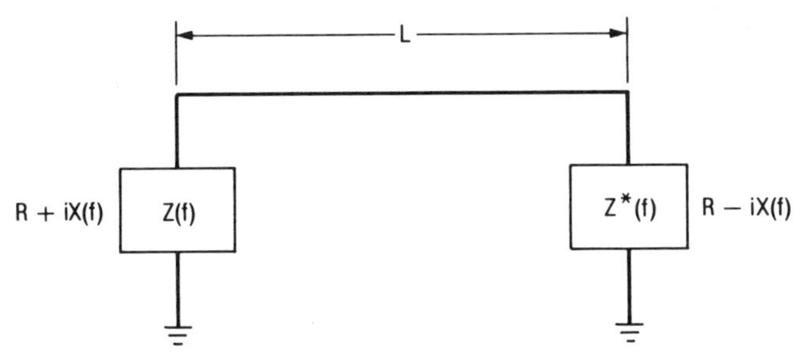

Figure 3.3

impedance of the receiver system, which is conjugate-matched to the antenna impedance for maximum signal-power transfer. In this case the noise power dissipated in $R(f)$, that is, the noise power at the input to the front-end amplifier, is

$$\frac{V^2(f)\,\Delta f}{4R(f)} = kT\,\Delta f \tag{3.19}$$

and by (3.12) the power spectral density for thermal noise is simply kT, a constant, and therefore flat over the frequency range of interest. We also define the two-sided power spectral density, $kT/2$, for dealing with negative frequencies that arise in Fourier analysis. In this case the noise power is $\frac{1}{2}kT \times 2\,\Delta f = kT\,\Delta f$, as before.

3.4 POWER SPECTRAL DENSITY OF SHOT NOISE†

In deriving the power spectral density for shot noise let us consider a stream of electrons, the effect of which is observed at the output of a unity-gain filter (see Section 4.5) with real impulse response $h(t)$ (see Figure 3.4). Each electron acts as an impulse on the filter, producing an output current $i(t)$ consisting of a sum of pulses

$$i(t) = q \sum_{m=1}^{N} h(t - \tau_m) \tag{3.20}$$

where τ_m are the random arrival times of the electrons at the filter, corresponding to the arrival times of photons at the detector, and $\int h(t)\,dt = 1$ so that each current pulse $h(t)$ accounts for one unit q of electron charge. The current $i(t)$ in (3.20) is defined with N a very large

† This and the following sections make use of a number of results in Sections (4.3) and (4.5) which are self contained and can be read out of sequence.

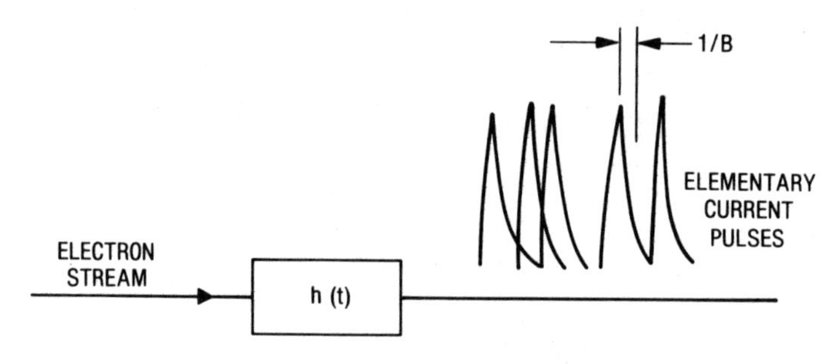

Figure 3.4 Shot noise model.

integer. The duration of each current pulse will be $\sim 1/B$ where B is the bandwidth of the filter $h(t)$ (see Section 4.3). As is discussed further in Ch. 5, shot noise is described by Poisson statistics. Referring to (2.11), if λ is the average number of electrons arriving per sec, then $\bar{\tau} = 1/\lambda$ is the average time between arrivals. The situation illustrated in Figure 3.4 is for $B \gg 1/\bar{\tau}$, thereby enabling individual current pulses to be resolved in time. If $B \ll 1/\bar{\tau}$ it is clear that there would be a great deal of overlap between successive pulses and the output would be smooth.

The shot-noise power spectral density will be derived by considering the mean-square current $\langle i^2(t) \rangle$, which is the average power dissipated in a 1-ohm resistor. Using (3.20) this is

$$\langle i^2(t) \rangle = \lim_{T \to \infty} \frac{1}{T} \int_{-T/2}^{T/2} i^2(t)\, dt = \lim_{T \to \infty} q^2 \sum_n \sum_m \frac{1}{T} \int_{-T/2}^{T/2} h(t - \tau_m) h(t - \tau_n)\, dt \tag{3.21}$$

Let

$$H(f) = \int_{-\infty}^{\infty} h(t) e^{-i2\pi f t}\, dt$$

and

$$\langle i^2 \rangle = \lim_{T \to \infty} q^2 \sum_n \sum_m \frac{1}{T} \int_{-T/2}^{T/2} dt \int_{-\infty}^{\infty} du \int_{-\infty}^{\infty} dv\, H(u) e^{-i2\pi u(t-\tau_n)} H(v) e^{-i2\pi v(t-\tau_m)}$$

$$= q^2 \sum_n \sum_m \frac{1}{T} \int_{-\infty}^{\infty} |H(f)|^2 e^{-i2\pi f(\tau_n - \tau_m)}\, df$$

where we have used

$$\lim_{T \to \infty} \int_{-T/2}^{T/2} e^{-i2\pi t(u+v)}\, dt = \delta(u + v) \tag{3.2.2}$$

where $\delta(x)$ is the Dirac delta function (see Section 4.5) and $H(u) = H^*(-u)$ since $h(t)$ is real.

Now, since the time τ_n required for n electrons to arrive at the filter (or for n photons to arrive at the focal plane of an optical detector) is given by: $\tau_n = n/\lambda$, then $(\tau_n - \tau_m) = (n - m)\bar{\tau}$ and

$$\sum_{n=1}^{N} \sum_{m=1}^{N} e^{-i2\pi f(\tau_n - \tau_m)} = \sum_{n=1}^{N} \sum_{m=1}^{N} e^{-i2\pi f\bar{\tau}(n-m)}$$

By making the change of variable $k = n - m$ this becomes

$$\sum_{k=1-N}^{0} \sum_{n=1}^{N+k} e^{-i2\pi kf\bar{\tau}} + \sum_{k=0}^{N-1} \sum_{n=k+1}^{N} e^{-i2\pi kf\bar{\tau}}$$

$$= N \sum_{k=-(N-1)}^{(N-1)} e^{-i2\pi kf\bar{\tau}} + \sum_{k=-(N-1)}^{0} ke^{-i2\pi kf\bar{\tau}} - \sum_{k=0}^{N-1} ke^{-i2\pi kf\bar{\tau}} \quad (3.23)$$

It is left as an exercise (see Exercise 3.18) to show that the summations involving the factor k vanish in the integration. Hence, by making use of the identity (see Exercise 3.3)

$$\sum_{k=-\infty}^{\infty} e^{-i2\pi kf\bar{\tau}} = \frac{1}{\bar{\tau}} \sum_{k=-\infty}^{\infty} \delta\left(f - \frac{k}{\bar{\tau}}\right) \quad (3.24)$$

it follows that as $N \to \infty$

$$q^2 \sum_n \sum_m \frac{1}{T} \int_{-\infty}^{\infty} |H(f)|^2 e^{-i2\pi f(\tau_n - \tau_m)}\, df = q^2 \frac{N}{T} \int_{-\infty}^{\infty} |H(f)|^2 \frac{1}{\bar{\tau}} \sum_k \delta\left(f - \frac{k}{\bar{\tau}}\right) df$$

$$= q^2 \frac{N}{T} \frac{1}{\bar{\tau}} \sum_k |H(k/\bar{\tau})|^2 \quad (3.25)$$

Referring to Figure 3.5, since the bandwidth B of $H(f)$ is large in comparison with $1/\bar{\tau}$ (3.25) can be written as

$$q^2 \frac{N}{T} \frac{|H(0)|^2}{\bar{\tau}} + q^2 \frac{N}{T} \frac{2}{\bar{\tau}} \sum_{k=1}^{B\bar{\tau}} |H(k/\bar{\tau})|^2$$

$$\sim q^2 \frac{N}{T} \frac{|H(0)|^2}{\bar{\tau}} + q^2 \frac{N}{T} \int_{-B}^{B} |H(f)|^2\, df \quad (3.26)$$

But $N/T \sim 1/\bar{\tau} = \lambda$ is the average number of electron arrivals per second, $|H(0)|^2 = 1$, and we recall that the two-sided noise bandwidth B of $H(f)$ is defined as

$$2B = \frac{1}{|H(0)|^2} \int_{-\infty}^{\infty} |H(f)|^2\, df \quad (3.27)$$

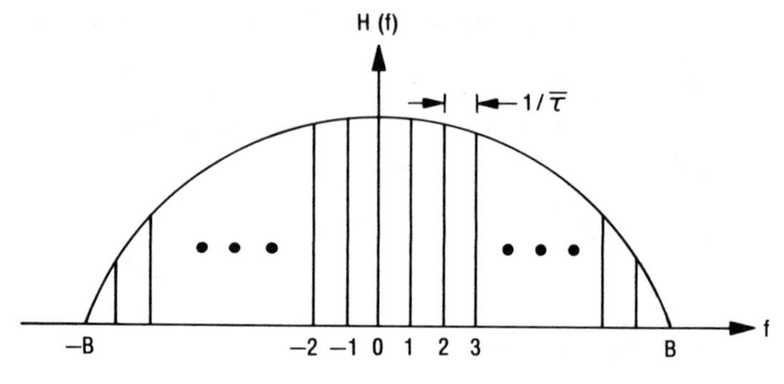

Figure 3.5

Therefore the first term in (3.26) is

$$q^2 \frac{N}{T} \frac{1}{\bar{\tau}} |H(0)|^2 = q^2\lambda^2 = I_{\mathrm{DC}}^2 \tag{3.28}$$

where I_{DC} is the DC current $q\lambda$, and the second term is

$$q^2 \frac{N}{T} \int_{-\infty}^{\infty} |H(f)|^2 \, df = 2BqI_{\mathrm{DC}}$$

Hence

$$\langle i^2(t) \rangle = I_{\mathrm{DC}}^2 + 2BqI_{\mathrm{DC}} \tag{3.29}$$

and the shot-noise power in a 1-ohm resistor, the variance of the shot noise current, is

$$2BqI_{\mathrm{DC}} = \langle i^2(t) \rangle - I_{\mathrm{DC}}^2 \tag{3.30}$$

The two-sided shot-noise power spectral density is therefore (3.12)

$$qI_{\mathrm{DC}} \tag{3.31}$$

which is white, as it is for thermal noise.

The shot noise becomes more severe as B increases. Suppose $B \ll 1/\bar{\tau}$. Then (see Exercise 3.3)

$$\begin{aligned} \langle i^2(t) \rangle &= q^2 \frac{N}{T} \int |H(f)|^2 \frac{1}{\bar{\tau}} \sum_k \delta\left(f - \frac{k}{\bar{\tau}}\right) df \\ &= q^2 \frac{N}{T} \frac{|H(0)|^2}{\bar{\tau}} = I_{\mathrm{DC}}^2 \end{aligned} \tag{3.32}$$

and the noise term vanishes. This takes place because the bandwidth of $H(f)$ is so small that all fluctuations due to random arrival times become smoothed out.

As in all such cases, the major consideration is the ratio of the standard deviation to the mean which in this case is $(2Bq/I_{DC})^{1/2}$. Hence, the fluctuations, effectively, become increasingly smoothed out as the average current increases.

3.5 SHOT NOISE AND OPTICAL RECEIVERS—THE QUANTUM LIMIT

Although laser radars are subject to thermal noise as well as shot noise, in coherent laser radars, which employ optical heterodyne detection, it is possible, if the optical local oscillator power is sufficiently large, to effectively eliminate the thermal noise, whence the ultimate receiver sensitivity, determined by the randomness in the production of photoelectrons, can be achieved.

A typical optical heterodyne receiver is illustrated in Figure 3.6. In such receivers the polarization alignment between the local oscillator and the signal of interest is a major consideration [11]. That is, it is essential that, after the beam splitter, the polarization vector of the transmitted signal and the reflected local oscillator signal be aligned to within a fraction of a wavelength over the detector surface. Specifically, if the mismatch in angle between the two vectors is α, we require that $D \sin \alpha \ll \lambda$ where D is the detector-surface width. Otherwise the interference between the two beams will not take place coherently over D. For example, taking an extreme case, if $D \sin \alpha \approx \lambda$, an interference null will occur somewhere on the detector.

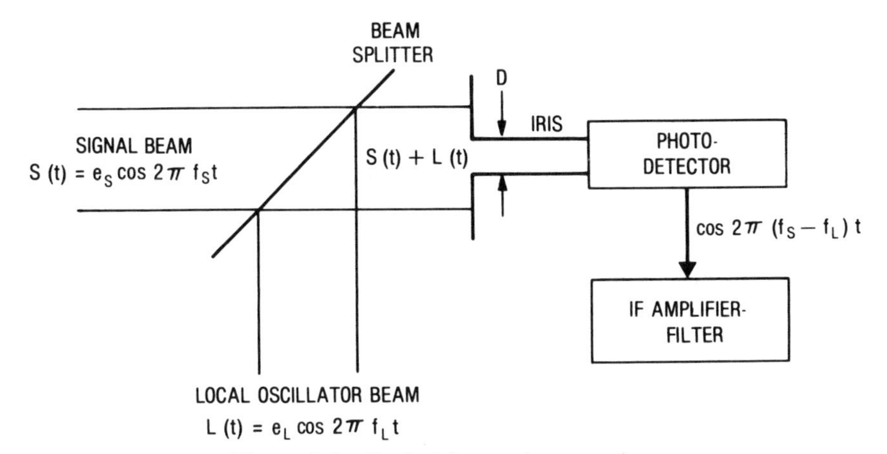

Figure 3.6 Optical heterodyne receiver.

The electronic current at the output of the optical detector is proportional to the power in the light signal and therefore proportional to the square of the total field consisting of signal plus local oscillator. Denote the signal and local-oscillator optical fields as $e_s \cos 2\pi f_s t$ and $e_L \cos 2\pi f_L t$, where the signal amplitude e_s is constant. Then

$$\begin{aligned} i = \kappa[e_s^2 \cos^2 2\pi f_s t + e_L^2 \cos^2 2\pi f_L t + e_L e_s \cos 2\pi(f_s + f_L)t \\ + e_L e_s \cos 2\pi(f_s - f_L)t] \end{aligned} \tag{3.33}$$

where κ is a constant defined below. The detector cannot respond to the optical frequencies f_s, f_L and $f_L + f_s$ which are of the order of 10^{14} Hz. It therefore produces current proportional to the mean-square values of these terms. On the other hand, the IF frequency $f_{IF} = f_s - f_L$ is generally of the order of $\sim 10^7$ Hz. Hence (3.33) becomes

$$\begin{aligned} i &= \kappa\left(\frac{e_s^2}{2} + \frac{e_L^2}{2} + e_L e_s \cos 2\pi f_{IF} t\right) = i_{DC} + i_{IF} \\ &= i_{DC}\left[1 + \frac{2e_L e_s}{e_s^2 + e_L^2} \cos 2\pi f_{IF} t\right] \approx i_{DC}\left[1 + \frac{2e_s}{e_L} \cos 2\pi f_{IF} t\right] \end{aligned} \tag{3.34}$$

if, as will be the case, $e_L \gg e_s$. The mean-square output current $\langle i_{IF}^2 \rangle$ of the IF amplifier is

$$\langle i_{IF}^2 \rangle = \frac{2e_s^2}{e_L^2} i_{DC}^2 = \frac{2P_s}{P_L} i_{DC}^2 \tag{3.35}$$

therefore, where P_s and P_L are the optical-signal and local-oscillator power, respectively.

Now, $i_{DC} = \kappa(P_s + P_L) \approx \kappa P_L$ and

$$\langle i_{IF}^2 \rangle = \frac{2P_s}{P_L} \kappa^2 P_L^2 = 2\kappa^2 P_s P_L \tag{3.36}$$

Hence the IF current which represents the input signal can, ideally, be increased to whatever value may be desired by increasing the local-oscillator power.

The noise current i_n at the output consists of shot noise plus thermal noise and the mean-square value is, from the foregoing results

$$\langle i_n^2 \rangle = 2q\kappa(P_s + P_L)B + \frac{kTB}{R} \sim 2qP_L \kappa B + \frac{kTB}{R} \tag{3.37}$$

where R is the effective load resistance and T the effective noise temperature, which includes the amplifier noise temperature—to be discussed below. The signal-to-noise ratio can therefore be defined in this case as

$$\mathrm{SNR} = \frac{2\kappa^2 P_s P_L}{2q\kappa P_L B + \dfrac{kTB}{R}} \tag{3.38}$$

and if the local-oscillator power is made sufficiently large—typically of the order of milliwatts—the shot-noise term dominates and (3.38) becomes

$$\mathrm{SNR} = \frac{\kappa P_s}{qB} \tag{3.39}$$

If N_p is the number of photons per second arriving at a detector and N_e is the number of electrons produced, the quantity κ is defined by the relationship

$$i = qN_e = \kappa h\nu N_p$$

where h is Planck's constant and ν the optical frequency. Hence

$$\kappa = \frac{q\eta}{h\nu}$$

where $\eta = N_e/N_p$ is the quantum efficiency of the detector. Equation (3.39) thus becomes

$$\mathrm{SNR} = \frac{\eta P_s}{h\nu B} \tag{3.40}$$

and the detection system is said to be quantum limited.

Since $P_s = N_p h\nu$, (3.40) becomes $\eta \hat{N}_p$ where $\hat{N}_p$ is the total number of photons in the signal, because $1/B$ is the nominal signal duration. Thus the minimum signal power $(P_s)_{\min}$—or equivalently the minimum number of photons $(\hat{N}_p)_{\min}$—required to produce a value of SNR equal to unity is

$$(P_s)_{\min} = \frac{h\nu B}{\eta} \tag{3.41}$$

$$(\hat{N}_p)_{\min} = \frac{1}{\eta}$$

being dependent only on the quantum efficiency of the detector when thermal noise is effectively eliminated.

For purposes of this illustration, in which the signal amplitude e_s is constant—as is therefore also P_s—the quantity in (3.40) is usually termed carried-to-noise ratio (CNR) rather than SNR, since in an actual application the signal amplitude could be fluctuating as described in Chapter 1. Equation (3.40) however is actually valid in either case for a heterodyne receiver, except that if e_s and therefore P_s exhibit random fluctuations, then SNR becomes a random variable [12].

Since the shot-noise fluctuations are inherent in the signal itself, (3.41) represents the fundamental limit on the weakest signal that an optical sensor is capable of detecting, and the ultimate sensitivity of optical sensors is achieved using heterodyne receivers. There is however another feature of such receivers. Let the signal and local-oscillator time functions be written as $e_s \cos(2\pi f_s t + \theta_s)$ and $e_L \cos(2\pi f_L t + \theta_L)$, where θ_s is unknown. The IF output is then of the form $\cos(2\pi f_{IF} t + \theta_s - \theta_L)$, and if θ_L is known then θ_s can be measured by measuring the phase of the electrical IF signal.

3.6 NOISE STATISTICS—SHOT NOISE IN RADAR AND LASER RADAR

By the law of large numbers and the central-limit theorem discussed in Chapter 2 it is clear that thermal noise has a Gaussian distribution given by

$$P(x) = \frac{1}{\sqrt{2\pi}\sigma} e^{-x^2/2\sigma^2} \tag{3.42}$$

where, assuming a properly matched impedance condition, $\sigma^2 = kTB$. That is, the observed value of a thermal noise voltage $n(t)$, for any t, is the sum of the contributions from a very large number of elementary electronic current pulses, and by the central limit theorem the numerical value of the random variable $n(t_0)$ for any time t_0 therefore has a probability density function given by (3.42). In terms of the foregoing discussion, we say that $n(t)$ is a realization of a Gaussian random process, and thermal noise is referred to as Gaussian noise.

For shot noise, fundamentally, Poisson statistics applies [13] with the probability of k events—electron emissions, photon arrivals, etc.—in T seconds given by

$$P(k, T) = \frac{(\lambda T)^k e^{-\lambda T}}{k!} \tag{3.43}$$

where λ is the average number of events per second. For λT sufficiently large, the Poisson distribution can be approximated by the Gaussian distribution as discussed in connection with Table (2.1). In either case, because shot noise is described by a single-parameter distribution, the ratio of standard deviation to the mean is $1/\sqrt{n}$ where $n = \lambda T$ is the number of photons that are received during an observation interval. This bears directly on the discussion in Chapter 1 concerning the relative importance of shot noise in radar and laser-radar systems. Since the energy per photon is $h\nu$, where ν is the carrier frequency, then the energy E in the signal is: $E = nh\nu$, and $1/\sqrt{n} = \sqrt{h\nu/E}$. But laser-radar frequencies exceed radio frequencies by a factor of $\sim 10^8$. Thus, all other things being equal, $1/\sqrt{n}$ for laser radar can exceed that for a microwave radar by a factor of $\sim 10^4$. Along with this, we have seen in the preceding section that thermal noise can be greatly reduced in laser-radar systems by means of coherent operation and, in addition, optical detectors can be made to be extremely quiet. As a result, it is meaningful in laser-radar systems to speak of, say, 5 photons per detection interval. On the other hand, in microwave radar $E/h\nu$ might have to be of the order of 10^6 in order for the signal to be observable in the presence of noise due to random currents. Looking ahead to Sections 5.5 and 6.3.1, this discussion can be summarized by noting that $(E/h\nu)^{1/2}$, the square-root of the number of photons received during an observation interval, is, effectively, the shot-noise signal-to-noise ratio. In laser radar other forms of noise can be made very low, and since the photon energy is so large, signal-to-noise ratios for other types of noise can be very large in comparison and shot noise therefore can be the dominant consideration. On the other hand, in radar, for the power levels required to overcome thermal noise the shot-noise signal-to-noise ratio is so large because of the relatively low energy per photon, that shot noise is not a consideration.

We now consider the Rayleigh and Rice distributions which were introduced in Chapter 2. Both deal with Gaussian noise at the output of a band-pass filter. Noise at the output of a band-pass filter has a very different character from noise at the output of a baseband, or video filter. As seen in Figure 3.7, baseband noise, which includes the frequency origin $f = 0$, or DC, appears simply as a function which varies randomly in time. Bandpass noise on the other hand has the appearance of a random amplitude modulation of a carrier whose phase also varies randomly in time. Bandpass noise can therefore be represented as

$$r(t)\cos(2\pi f_0 t + \phi(t)) = n_c(t)\cos 2\pi f_0 t + n_s(t)\sin 2\pi f_0 t \qquad (3.44)$$

The representation for bandpass noise on the right-hand side of (3.44) although introduced by Rice [14] was, according to Rice, actually first considered by Einstein and Von Laue in 1911–1915 [15], who showed that $n_c(t)$ and $n_s(t)$ are Gaussian with the same mean and variance, are independent, and that $\phi(t)$ for any value of t is uniformly distributed over $(0, 2\pi)$.

The rate at which $r(t)$ and $\phi(t)$ vary in time depends on the bandwidth B of the bandpass filter. Because in the early days of radar the fractional bandwidths B/f_0 were quite small, that is $B \ll f_0$, (3.44) has traditionally been referred to as the narrow-band representation for Gaussian noise. However there is really no need for such a restriction. As has been shown by Viterbi [16], it is only necessary that, as illustrated in Figure 3.8, the positive (and negative) bandpass characteristics do not extend below (or above) the frequency origin. This means that we must have, nominally,

a. WIDEBAND BASEBAND NOISE

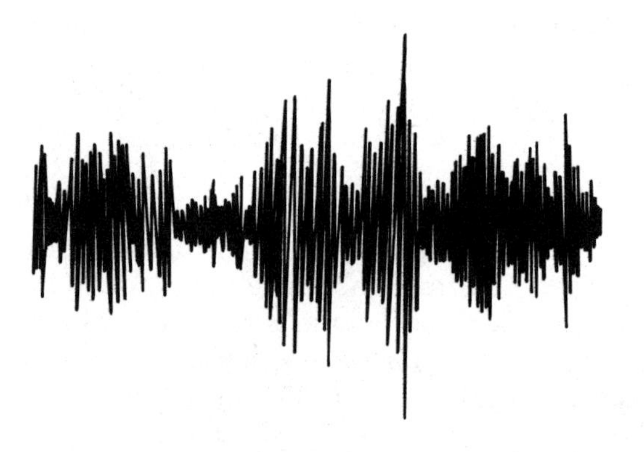

b. AFTER PASSAGE THROUGH A NARROW BANDPASS FILTER

Figure 3.7 Baseband and bandpass Gaussian noise.

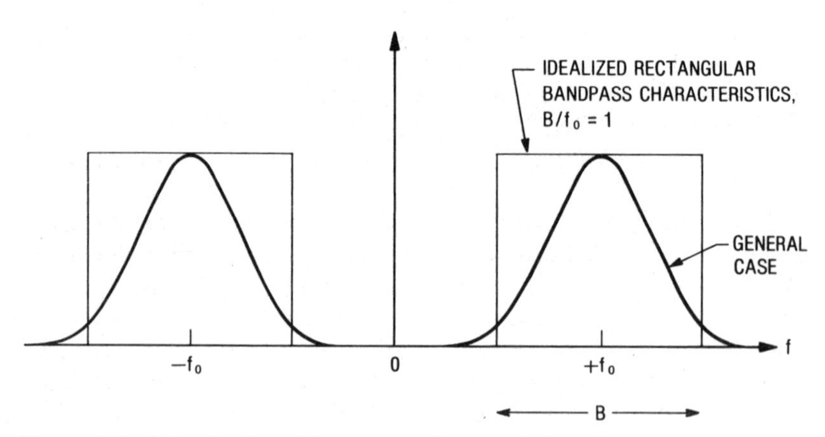

Figure 3.8 Magnitudes of frequency characteristics of bandpass functions.

$f_0 \geq B/2$. But since the duration T of a pulse of bandwidth B is nominally $T \sim 1/B$ (see Section 4.3), the restriction translates to $Tf_0 \geq \frac{1}{2}$, or that there be at least $\frac{1}{2}$ cycles of the carrier in each pulse. But the notion of an amplitude-modulated carrier essentially begins to lose meaning if there is less than, say, at least one cycle of carrier in the duration of the signal. Therefore, with the assumption of amplitude-modulated time functions no other restrictions are necessary.

As is discussed in Chapter 5, for certain types of detection schemes the magnitude $r(t)$ is of interest, which is also referred to as the envelope of the output of the bandpass filter. Since envelope detection is a nonlinear operation the result is no longer Gaussian. Let

$$n(t) = n_c(t) \cos 2\pi f_0 t + n_s(t) \sin 2\pi f_0 t \tag{3.45}$$

$$\text{with } E[n(t)] = 0,\ E[n^2(t)] = \sigma^2.$$

Then
$$E[n^2(t)] = E[n_c^2(t)] \cos^2 2\pi f_0 t + E[n_s^2(t)] \sin^2 2\pi f_0 t + 2E[n_c(t) n_s(t)] \sin 2\pi f_0 t \cos 2\pi f_0 t = \sigma^2 \tag{3.46}$$

But since $E[n_c(t)] = E[n_s(t)] = 0$, and $n_c(t)$ and $n_s(t)$ are independent, then $E[n_c(t) n_s(t)] = 0$, and $E[n^2(t)] = E[n_c^2(t)] = E[n_s^2(t)] = \sigma^2$. Now for any value of t the two independent random variables $n_c(t)$ and $n_s(t)$ can be represented as taking values on the x and y axes respectively in a plane (see Figure 3.9) with $x = n_c = r \cos \phi$ and $y = n_s = r \sin \phi$, and $dx\, dy = r\, dr\, d\phi$.

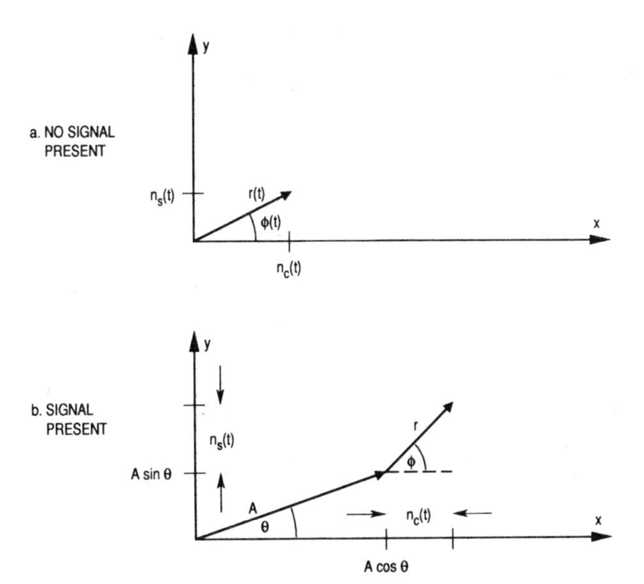

Figure 3.9 In-phase and quadrature representation of noise with and without signal.

And since they are independent and Gaussian, then

$$P(r, \phi)\,dr\,d\phi = P(x, y)\,dx\,dy = e^{-x^2/2\sigma^2} e^{-y^2/2\sigma^2} \frac{dx\,dy}{2\pi\sigma^2}$$

$$= e^{-r^2/2\sigma^2} \frac{r\,dr\,d\phi}{2\pi\sigma^2} = P(r)\,dr\,P(\phi)\,d\phi \tag{3.47}$$

which proves that r and ϕ are independent, and

$$P(r) = \frac{re^{-r^2/2\sigma^2}}{\sigma^2}$$

$$P(\phi) = \frac{1}{2\pi}, \quad 0 \leq \phi \leq 2\pi \tag{3.48}$$

Equation (3.48) is the Rayleigh distribution illustrated in Figure 3.10. Here $E(r) = \sqrt{\frac{1}{2}\pi}\sigma$, $E(r^2) = 2\sigma^2$ and $\mathrm{Var}(r) = \sigma^2(4 - \pi)/2 = 0.43\sigma^2$.

The Rayleigh distribution applies to noise alone. The Rice distribution applies to the statistics of the envelope of the output of a bandpass filter consisting of signal plus noise in the form (Figure 3.9b):

$$s(t) + n(t) = A\cos(2\pi f_0 t + \theta) + n_c(t)\cos 2\pi f_0 t + n_s(t)\sin 2\pi f_0 t$$

$$= (n_c(t) + A\cos\theta)\cos 2\pi f_0 t + (n_s(t) - A\sin\theta)\sin 2\pi f_0 t \tag{3.49}$$

where θ is the unknown signal phase. In this case (3.47) becomes

$$P(r, \phi)\,dr\,d\phi = \exp\left[-\frac{(x + A\cos\theta)^2}{2\sigma^2}\right]\exp\left[-\frac{(y - A\sin\theta)^2}{2\sigma^2}\right]\frac{dx\,dy}{2\pi\sigma^2}$$

$$= \exp\left(-\frac{r^2 + A^2}{2\sigma^2}\right)\exp\left[-\frac{Ar\cos(\phi + \theta)}{\sigma^2}\right]\frac{r\,dr\,d\phi}{2\pi\sigma^2} \tag{3.50}$$

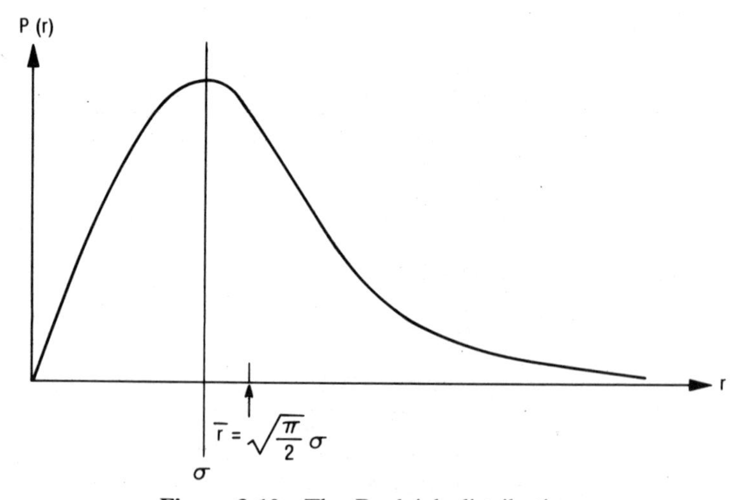

Figure 3.10 The Rayleigh distribution.

and, unlike the Rayleigh distribution, $P(r, \phi)$ cannot be factored into $P(r, \phi) = P(r) \times P(\phi)$ and r and ϕ and are therefore not independent. In order to determine $P(r)$ it is therefore necessary to use

$$P(r) = \int_0^{2\pi} P(r, \phi)\, d\phi = \exp\left(-\frac{r^2 + A^2}{2\sigma^2}\right) \frac{r}{2\pi\sigma^2} \int_0^{2\pi} \exp\left(\frac{-Ar\cos\phi}{\sigma^2}\right) d\phi \tag{3.51}$$

where θ vanishes with a change of variables, and by making use of the identity

$$\frac{1}{2\pi} \int_0^{2\pi} \exp\left(-\frac{Ar\cos\phi}{\sigma^2}\right) d\phi = I_0\left(\frac{Ar}{\sigma^2}\right) \tag{3.52}$$

where $I_0(x)$ is the zero-order modified Bessel function of the first kind, equation (3.51) yields the Rice distribution

$$P(r) = \frac{r}{\sigma^2} \exp\left(-\frac{r^2 + A^2}{2\sigma^2}\right) I_0\left(\frac{rA}{\sigma^2}\right) \tag{3.53}$$

which describes the statistics of the magnitude (or envelope) of the output of a bandpass filter consisting of sinusoidal signal plus noise.

The Rice distribution has the appearance illustrated in Figure 3.11 and, in fact, for large signal-to-noise ratio $A^2/2\sigma^2$, it approaches a Gaussian distribution with mean A and variance σ^2. The Rayleigh and Rice distributions will be utilized extensively in Chapter 6 which deals with coherent and noncoherent detection.

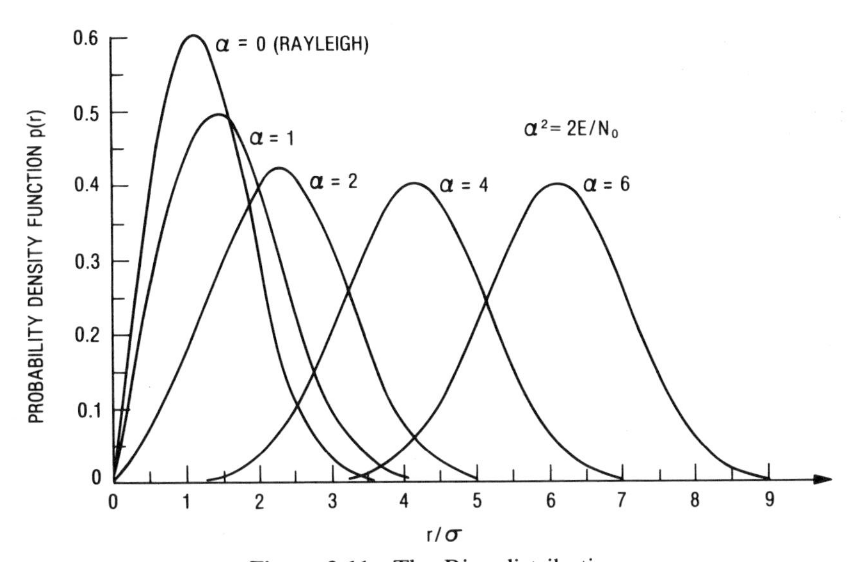

Figure 3.11 The Rice distribution.

3.7 NOISE FIGURE AND NOISE TEMPERATURE

Nyquist's theorem shows that thermal noise depends on the system temperature, and the effective noise temperature of a system is therefore a measure of the thermal noise produced by the system. Consider Figure 3.12, in which, conceptually, the noisy amplifier with gain G is replaced by a noiseless amplifier and a noise source at temperature T_A; the input impedance of the amplifier is R. Then $P_0 = kT_A BG$ and the amplifier noise temperature T_A is defined as

$$T_A = \frac{P_0}{kBG} \tag{3.54}$$

The noise figure F of an amplifier is related to the noise temperature and is defined as

$$F = \frac{(S/N)_{in}}{(S/N)_{out}} \tag{3.55}$$

Noise figure is measured as shown in Figure 3.13. The input is a standard noise source with power kT_0B where B is the noise bandwidth of the amplifier (3.27) and $T_0 = 290$ K. Note that F is independent of the load impedance.

Now since $S_{out} = GS_{in}$ we have from (3.55)

$$F = \frac{S_{in}}{S_{out}} \frac{N_{out}}{N_{in}} = \frac{1}{G} \frac{GN_{in} + N_{amp}}{N_{in}} = 1 + \frac{N_{amp}}{GkT_0B} \tag{3.56}$$

where from (3.54) $N_{amp} = GkT_A B$. Hence

$$G(F-1)kT_0B = N_{amp} = GkT_A B \tag{3.57}$$

and

$$F = 1 + \frac{T_A}{T_0} \quad \text{or} \qquad T_A = T_0(F-1) \tag{3.58}$$

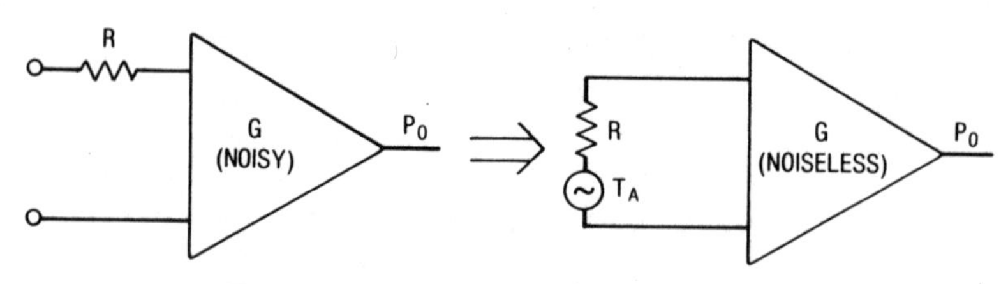

Figure 3.12 Model of a noisy amplifier.

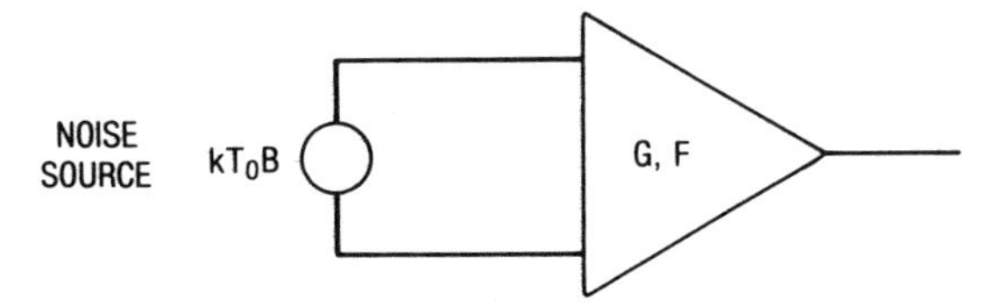

Figure 3.13 Measurement of noise figure.

Thus a noise figure of 1, or 0 dB, represents an ideal noiseless amplifier.

The situation of general interest is that shown in Figure 3.14, in which the noise figure and noise temperature of a system consisting of a number of amplifiers in series is to be determined. Setting $B = 1$ for convenience, yields $N_1 = G_1kT_0 + G_1kT_1 = G_1kT_0 + G_1kT_0(F_1 - 1)$, $N_2 = G_2N_1 + G_2kT_0(F_2 - 1)$ etc., and it is easily shown using (3.56) that the composite noise figure of the system is

$$F_{123} = F_1 + \frac{F_2 - 1}{G_1} + \frac{F_3 - 1}{G_1G_2} + \text{etc.} \tag{3.59}$$

or equivalently the total noise temperature T_{123} is

$$T_{123} = T_1 + \frac{T_2}{G_1} + \frac{T_3}{G_1G_2} + \frac{T_4}{G_1G_2G_3} + \text{etc.} \tag{3.60}$$

From (3.59) and (3.60) the first, front-end, stage of amplification determines the noisiness of the system, assuming of course that G_1 is large enough. Thus, if the front-end amplifier is of high quality (low value of F) the amplifiers in succeeding stages can be of lower quality with negligible system noise degradation.

Equation (3.60) represents the noise generated in receiver system components. The total system noise temperature is obtained by adding to this the antenna temperature, which represents ambient background noise measured at the antenna output terminals—before the front-end amplifier—with a load impedance conjugate matched to the antenna impedance.

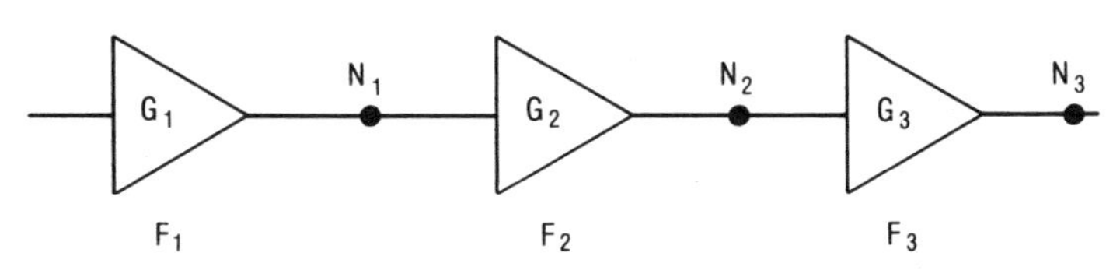

Figure 3.14 Noisy amplifiers in series.

3.8 NOISE FIGURE OF AN ATTENUATOR

Lossy elements can be looked at as noisy amplifiers with gain less than unity. As remarked above in Section 3.1 only lossy elements contribute noise to the system.

Consider Figure 3.15 in which the attenuator is in thermal equilibrium with a heat reservoir at arbitrary temperature T_L. The net power incident on the attenuator is kT_L (letting $B = 1$) and the power in the cavity is kT_L/L. But in thermal equilibrium the power absorbed by the attenuator must be equal to the power emitted by the attenuator kT_A/L, where T_A is the effective noise temperature of the attenuator. Hence

$$kT_L - \frac{kT_L}{L} = kT_L\left(\frac{L-1}{L}\right) = \frac{kT_A}{L} \tag{3.61}$$

or

$$T_A = T_L(L-1) \tag{3.62}$$

and the noise temperature of a lossy element is its actual temperature multiplied by $L - 1$, and is therefore zero if $L = 1$.

Now referring to (3.56)

$$F = \frac{1}{G}\,\frac{N_{out}}{N_{in}} = L\left(\frac{\dfrac{kT_0}{L} + \dfrac{kT_A}{L}}{kT_0}\right) = 1 + \frac{T_L}{T_0}(L-1) \tag{3.63}$$

where T_L is the temperature of the lossy element. If $T_L = 290$ K, then $F = L$.

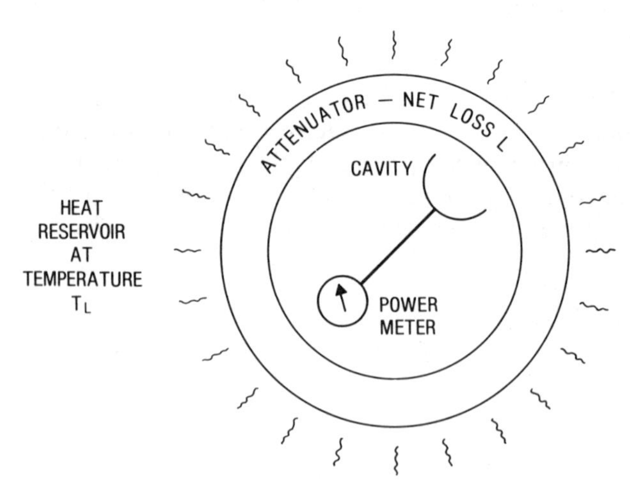

Figure 3.15 Cavity in thermal equilibrium with heat reservoir at temperature T_L.

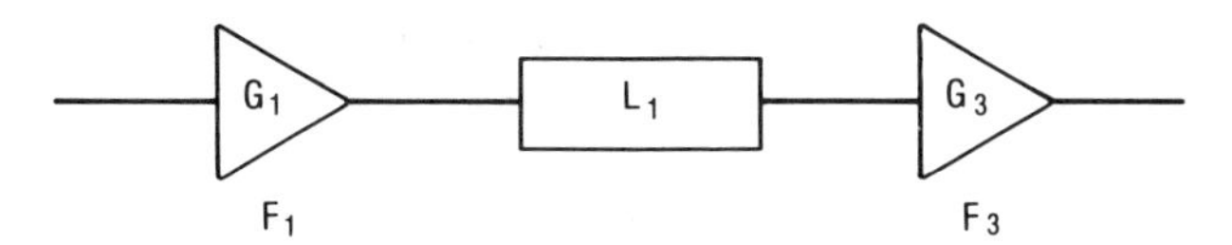

Figure 3.16 Amplifiers connected by lossy transmission line.

As an example consider the situation in Figure 3.16 in which the lossy element is at room temperature, 290 K. Referring to (3.59) with $G_2 = 1/L$ and $F_2 = L$

$$F_{123} = F_1 + \frac{(L-1)}{G_1} + \frac{L(F_3 - 1)}{G_1} \tag{3.64}$$

and the lossiness of the attenuator, which might represent the cable from a TV antenna to the TV set, is effectively reduced by G_1. It is for this reason that TV booster amplifiers are located at the antenna, before the cable. That is, there is no point in amplifying the noise generated in the transmission path between the antenna and the set. By using the results of Sections 3.7 and 3.8, the total noise figure of a receiver system consisting of noisy amplifiers connected by lossy elements can be calculated.

3.9 APPLICATIONS: NOISE POWER MEASUREMENTS

The average power of a random process with realization $x(t)$ is

$$P_{\text{av}} = \lim_{T\to\infty} \frac{1}{T} \int_{-T/2}^{T/2} x^2(t)\, dt \tag{3.65}$$

Since all measurements must be of finite duration, then any power measurement is necessarily a random variable given by

$$P(T) = \frac{1}{T} \int_{-T/2}^{T/2} x^2(t)\, dt \tag{3.66}$$

where the random variable $P(T)$ is dependent on T. As $T \to \infty$ $P(T)$ approaches a constant value given by (3.65). If $x(t)$ is ergodic then

$$\begin{aligned} E[P(T)] &= \frac{1}{T} \int_{-T/2}^{T/2} E[x^2(t)]\, dt = E[x^2(t)] \\ &= \lim_{T\to\infty} \frac{1}{T} \int_{-T/2}^{T/2} x^2(t)\, dt = P_{\text{av}} \end{aligned} \tag{3.67}$$

Suppose $x(t)$ is a realization of thermal noise measured at the output of a low-pass filter of bandwidth B. The question arises concerning how long the integration time T must be so that the standard deviation of the random variable $P(T)$ is some fraction K of its mean value $N_0 B$ where N_0 is the one-sided noise power spectral density (see (3.10) and (3.83)). If K is not sufficiently small the measurement will be of little value since the fluctuations in the random quantity $P(T)$ will be too large for it to be a good representation of P_{av}. To determine this we write

$$\text{Var}[P(T)] = \frac{1}{T^2} E \int_{-T/2}^{T/2} \int_{-T/2}^{T/2} [x^2(t_1)x^2(t_2)] \, dt_1 \, dt_2 - [E[x^2(t)]]^2 \tag{3.68}$$

and using (3.14)

$$E[x^2(t_1)x^2(t_2)] = 2r^2(t_2 - t_1) + [r^2(0)]^2$$

This yields

$$\text{Var } P(T) = \frac{2}{T^2} \int_{-T/2}^{T/2} \int_{-T/2}^{T2} r^2(t_2 - t_1) \, dt_1 \, dt_2 \tag{3.69}$$

since $r(0) = E[x^2(t)]$. Now

$$\begin{aligned} r(t_2 - t_1) &= \int_{-\infty}^{\infty} W(f) e^{i2\pi f(t_2 - t_1)} = \frac{N_0}{2} \int_{-B}^{B} e^{i2\pi f(t_2 - t_1)} \, df \\ &= \frac{N_0}{2} \frac{\sin 2\pi B(t_2 - t_1)}{\pi(t_2 - t_1)} \end{aligned} \tag{3.70}$$

So

$$\text{Var } P(T) = \frac{N_0^2}{2T^2} \int_{-T/2}^{T/2} \int_{-T/2}^{T/2} \left[\frac{\sin 2\pi B(t_2 - t_1)}{\pi(t_2 - t_1)} \right]^2 dt_1 \, dt_2 \tag{3.71}$$

and if $T \gg 1/B$ then to a good approximation (see Exercise 3.4)

$$\text{Var}[P(T)] = \frac{N_0^2}{2T} \int_{-T/2}^{T/2} \left(\frac{\sin 2\pi B\tau}{\pi\tau} \right)^2 d\tau \sim \frac{N_0^2 B}{T} \tag{3.72}$$

since

$$\int_{-\infty}^{\infty} \left(\frac{\sin 2\pi B\tau}{\pi\tau} \right)^2 d\tau = 2B$$

Hence

$$[\text{Var } P(T)]^{1/2} \approx \left(\frac{N_0^2 B^2}{BT} \right)^{1/2} = \frac{P_{av}}{\sqrt{BT}} \tag{3.73}$$

This result is a practical example of (2.37) which shows that the standard deviation of the sum of n independent random variables goes as $\sim 1\sqrt{n}$. In this example, the noise, which is bandlimited to the range $|f| < B$, if viewed on an oscilloscope would exhibit a noise spike roughly every $1/B$ seconds, which illustrates the fact that noise samples separated by $1/B$ seconds are independent. This is dealt with in some detail in Chapter 5. Thus, in T seconds BT such spikes would be observed or, equivalently, during the time T there is an average over BT independent random events. Hence, the standard deviation of the random quantity being averaged becomes reduced by the square root of this number.

Noise power also can be estimated from spectral measurements. Consider

$$\int_{-B}^{B} |X_T(f)|^2 \, df \tag{3.74}$$

where

$$X_T(f) = \int_{-T/2}^{T/2} x(t) e^{-i2\pi f t} \, dt \tag{3.75}$$

Equation (3.74) can be written as

$$\begin{aligned} &\int_{-B}^{B} \int_{-T/2}^{T/2} x(t_1) e^{-i2\pi f t_1} \, dt_1 \int_{-T/2}^{T/2} x(t_2) e^{i2\pi f t_2} \, dt_2 \, df \\ &= \int_{-T/2}^{T/2} \int_{-T/2}^{T/2} x(t_1) x(t_2) \int_{-B}^{B} e^{-i2\pi f (t_1 - t_2)} \, df \, dt_1 \, dt_2 \\ &= \int_{-T/2}^{T/2} \int_{-T/2}^{T/2} x(t_1) x(t_2) \frac{\sin 2\pi B (t_1 - t_2)}{\pi (t_1 - t_2)} \, dt_1 \, dt_2 \end{aligned} \tag{3.76}$$

If $BT \gg 1$, then, since $\sim 90\%$ of the area under $(\sin(2\pi Bt)/\pi t$ is contained within the range $-1/B \leq t \leq 1/B$, the area under $[\sin 2\pi B(t_1 - t_2)]/\pi(t_1 - t_2)$ will be essentially contained within the range $-T/2 \leq (t_1, t_2) \leq T/2$ for all values of t_1 and t_2 except for a narrow border† of width $\sqrt{2}/B$. Therefore, since

$$\int_{-\infty}^{\infty} \frac{\sin 2\pi B t}{\pi t} \, dt = 1 \tag{3.77}$$

the function $[\sin 2\pi B(t_1 - t_2)]/\pi(t_1 - t_2)$ can under these conditions‡ to a

† The ratio of excluded area to total integration area T^2 is $\sim 1/BT$ which will be negligible under these conditions.

‡ Note that it is important that $x(t)$ be bandlimited so that it does not vary appreciably over time durations of the order of $1/B$.

good approximation be treated as the Dirac delta function $\delta(t_1 - t_2)$ (see Section 4.5), and (3.74) becomes

$$\int_{-B}^{B} |X_T(f)^2 \, df \approx \int_{-T/2}^{T/2} \int_{-T/2}^{T/2} x(t_1)x(t_2)\delta(t_1 - t_2) \, dt_1 \, dt_2 = \int_{T/2}^{T/2} x^2(t) \, dt \tag{3.78}$$

for large BT. Equation (3.78) is a practical extension of Parseval's theorem to finite ranges of time and frequency; Parseval's theorem is

$$\int_{-\infty}^{\infty} |X(f)|^2 \, df = \int_{-\infty}^{\infty} |x(t)|^2 \, dt \tag{3.79}$$

(See Exercise 3.15.) Therefore, using (3.66), the average power can be estimated from

$$P_{\text{av}} = \frac{1}{T} \int_{-B}^{B} |X_T(f)|^2 \, df \tag{3.80}$$

which is an approximation to (3.10) with $W(f)$ replaced by $|X_T(f)|^2/T$.

There are a number of useful definitions of signal-to-noise ratio SNR, some of which will be introduced in Chapters 5 and 6. One such definition, the integrated signal-to-noise ratio, $(\text{SNR})_\text{I}$, is as follows. Let the output of the low-pass filter discussed above in connection with (3.68) et seq. also contain a signal $s(t)$ with bandwidth $B' \le B$. Then

$$(\text{SNR})_\text{I} = \frac{\int_{-B}^{B} |S(f)|^2 \, df}{\int_{-B}^{B} |X(f)|^2 \, df} \tag{3.81}$$

where $S(f)$ is the Fourier transform of $s(t)$. But by (3.79), (3.78) and (3.66) this can be written as, assuming BT sufficiently large

$$(\text{SNR})_\text{I} = \frac{E}{TP(T)} \tag{3.82}$$

where $E = \int_{-\infty}^{\infty} s^2(t) \, dt$ is the signal energy. This can be put into a more familiar form as follows. For thermal noise, it is customary to denote the (one-sided) power spectral density kT of (3.19) as

$$kT = N_0 \tag{3.83}$$

Therefore, if we assume (3.66) to be a sufficiently good approximation to P_{av} which, referring to (3.10) is given by

$$P_{\text{av}} = r(0) = N_0 B \tag{3.84}$$

Then (3.82) becomes

$$(\mathrm{SNR})_{\mathrm{I}} = \frac{E}{TP_{\mathrm{av}}} = \frac{E}{BTN_0} \tag{3.85}$$

The ratio of signal energy to noise spectral density E/N_0, which can also be written as E/kT, is a very important quantity which will be seen to appear repeatedly throughout the text. Equation (3.85) provides a useful scaling between $(\mathrm{SNR})_{\mathrm{I}}$ and E/N_0. Also, if the noise in $(\mathrm{SNR})_{\mathrm{I}}$ is not white, (3.85) defines a white-noise process equivalent to the noise in (3.81). Note that if $|S(f)|^2$ and $|N(f)|^2$ are reasonably flat over the band of interest, then $(\mathrm{SNR})_{\mathrm{I}}$ is, approximately, just the ratio of the two spectral levels.

3.10 CONNECTIONS WITH STATISTICAL PHYSICS

A molecule of a gas is said to possess three "degrees of freedom," one for each independent spatial coordinate, and a volume of a gas consisting of N molecules possesses $3N$ degrees of freedom. For an ideal gas there are no intermolecular forces, and therefore no potential energy, and all the energy in the system is therefore contained in the form of kinetic energy. If the three components of molecular velocity are v_x, v_y and v_z, a fundamental result of statistical mechanics, the equipartition theorem, states that

$$\frac{1}{2}\,m\overline{v_x^2} = \frac{1}{2}\,m\overline{v_y^2} = \frac{1}{2}\,m\overline{v_z^2} = \frac{kT}{2} \tag{3.86}$$

where T is the temperature of the gas, assumed to be in thermal equilibrium, and k is Boltzmann's constant. Thus the energy of an ideal gas is stored in the form of kinetic energy and the energy per degree of freedom is $kT/2$; the total energy is of course $3NkT/2$.

This result also applies to electric circuits driven by thermal noise, for which the random motion of the electrons is exactly analogous to the random motion of gas molecules. In this case however, with the motion of the charged particles there arise magnetic as well as electric fields, and consequently stored electric and magnetic energy, whose equilibrium value is also prescribed by the equipartition theorem. We shall now prove this, which demonstrates the consistency of the foregoing results with physical laws.

As illustrated in Figure 3.17, let a resistor R at temperature T be in series with an inductor L. As is discussed in Section 3.2, the resistor can be viewed as a thermal noise source in series with R, for which the two-sided voltage spectral density is $2kTR$ (see (3.18)). If the random voltage $v(t)$ produces a random current $i(t)$ we can write

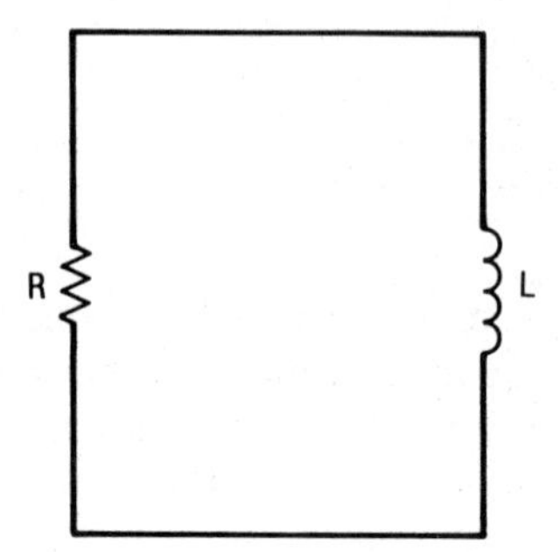

Figure 3.17

$$v(t) = L\,\frac{di(t)}{dt} + Ri(t) \tag{3.87}$$

The homogeneous solution is $i(t) = i_0 e^{-Rt/L}$, where $i(0) = i_0$. To determine the complete solution let

$$i(t) = C(t)e^{-Rt/L} \tag{3.88}$$

By substituting (3.88) into (3.87)

$$C(t) = \frac{1}{L}\int_0^t v(\tau)e^{R\tau/L}\,d\tau + i_0 \tag{3.89}$$

and

$$i(t) = i_0 e^{-Rt/L} + \frac{e^{-Rt/L}}{L}\int_0^t v(\tau)e^{R\tau/L}\,d\tau \tag{3.90}$$

Hence

$$E[i^2(t)] = i_0^2 e^{-2Rt/L} + \frac{e^{-2Rt/L}}{L^2}\int_0^t d\tau_1 \int_0^t d\tau_2\, e^{(\tau_1+\tau_2)R/L} E[v(\tau_1)v(\tau_2)] \tag{3.91}$$

Since the voltage power-spectral density is $2kTR$, then with the assumption of white noise (see Exercise 3.16).

$$E[v(\tau_1)v(\tau_2)] = 2kTR\delta(\tau_1 - \tau_2)$$

and

$$\begin{aligned} E[i^2(t)] &= i_0^2 e^{-2Rt/L} + e^{-2Rt/L}\,\frac{2kTR}{L^2}\int_0^t e^{2R\tau_1/L}\,d\tau_1 \\ &= \frac{kT}{L} + \left(i_0^2 - \frac{kT}{L}\right)e^{-2Rt/L} \end{aligned} \tag{3.92}$$

But the magnetic energy stored in L is $\frac{1}{2}L\overline{i^2}$ and therefore in equilibrium, as $t \rightarrow \infty$,

$$\frac{1}{2}L\overline{i^2} = \frac{kT}{2} \tag{3.93}$$

The identical result holds for the stored electrical energy in the circuit capacitance (see Exercise 3.12).

EXERCISES FOR CHAPTER 3

3.1 Show that the random process $\cos(2\pi ft + \theta)$, with θ uniformly distributed over $(0, 2\pi)$ is stationary, but does not satisfy (3.4).

3.2 Show that the process of Exercise 3.1 is ergodic.

3.3 Consider the series $d(t) = \Sigma_{n=-\infty}^{\infty} \delta(t - nT)$ where $\delta(t)$ is the Dirac delta function, which has the properties

$$\int_{\infty}^{\infty} \delta(t)\, dt = 1$$

$$\int_{-\infty}^{-\infty} f(t - \tau)\delta(t)\, dt = f(\tau)$$

The series is a periodic function which therefore can be written as

$$d(t) = \sum_n \delta(t - nT) = \sum_n C_n \exp\left(-i2\pi \frac{nt}{T}\right)$$

where

$$C_n = \frac{1}{T}\int_{-T/2}^{T/2} \delta(t) \exp\left(i2\pi \frac{nt}{T}\right) dt$$

Show that this leads to the important identity

$$\sum_{n=-\infty}^{\infty} \exp\left(-i2\pi \frac{nt}{T}\right) = T \sum_{n=-\infty}^{\infty} \delta(t - nT)$$

3.4 Prove the Wiener Khinchine theorem, (3.8) by using (3.7) and making an appropriate change of variables in the integral which alters the region of integration, and letting $T \rightarrow \infty$. This exercise also applies to (3.71) and (3.72).

3.5 Prove (3.11) and (3.12). Section (4.5) may be helpful.

3.6 White noise with power spectral density $N_0/2$ is input to filters with the following characteristics. What is the output power in each case?

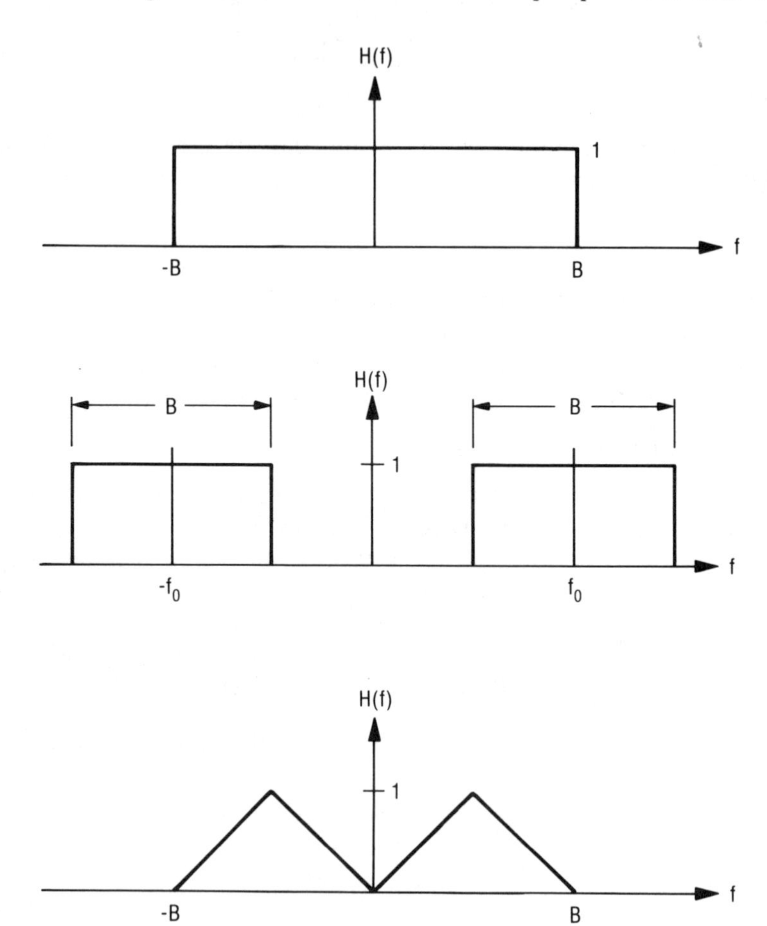

3.7 Fill in all missing steps in the derivation of Nyquist's theorem.

3.8 Using the representation on the left-hand side of (3.44), show that if $r(t)$ and $\phi(t)$ are independent, it then follows that $n_s(t)$ and $n_c(t)$ are independent, also that $E[n_s^2(t)] = E[n_c^2(t)]$.

3.9 Verify the sketch of the Rayleigh distribution in Figure 3.8 and the values of its moments.

3.10 By using the approximation $I_0(x) \sim e^x/\sqrt{2\pi x}$ for large x, show that the Rice distribution approaches a Gaussian distribution for large signal-to-noise ratio. Show that this approximation holds in the vicinity of the peak, but not necessarily in the tails of the distribution.

3.11 Verify (3.59) and (3.60) for three stages.

3.12 A resistor R at temperature T can also be viewed as thermal a current source, with current spectral density $2kT/R$ in parallel with a resistor R. Consider the situation illustrated in the accompanying figure. The random current will give rise to a mean square equilibrium voltage $\overline{v^2}$ across C (and R), and the equilibrium value of electric energy stored in C will be $\frac{1}{2}C\overline{v^2}$. Establish the above result for the current power spectral density using Nyquist's theorem and show that $\frac{1}{2}C\overline{v^2} = \frac{1}{2}kT$.

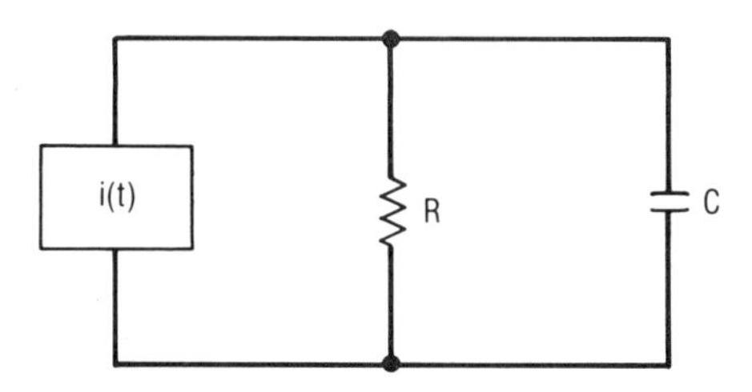

3.13 Let $x(t)$ be a realization of a random process with derivative defined in the usual way:

$$\frac{dx(t)}{dt} = \lim_{\Delta t \to 0} \frac{x(t + \Delta t) - x(t)}{\Delta t}$$

Find $E[dx(t)/dt]$. Show that the cross-correlation between the function and its derivative satisfies

$$E\left[x(t)\,\frac{dx(t)}{dt}\right] = \left.\frac{dr(\tau)}{d\tau}\right|_{\tau=0}$$

$$E\left[x(t)\,\frac{d}{d\tau}\,x(t + \tau)\right] = \frac{dr(\tau)}{d\tau}$$

where $r(\tau) = E[x(t)x(t + \tau)]$ and that the autocorrelation function of the derivative is

$$E\left[\frac{dx(t)}{dt}\,\frac{dx(t + \tau)}{dt}\right] = -\,\frac{d^2r(\tau)}{d\tau^2}$$

3.14 Determine which receiver is better, and calculate the output (S/N).

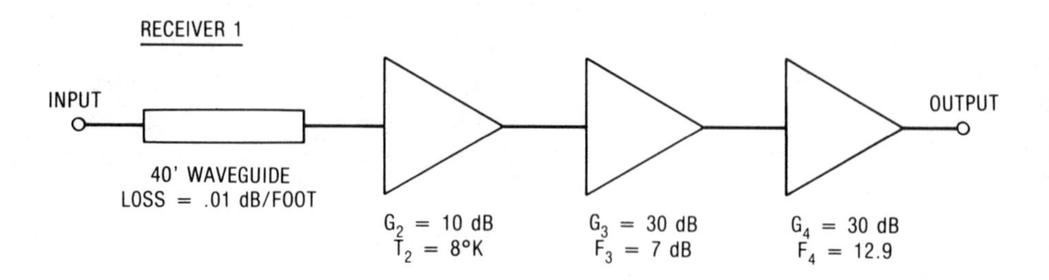

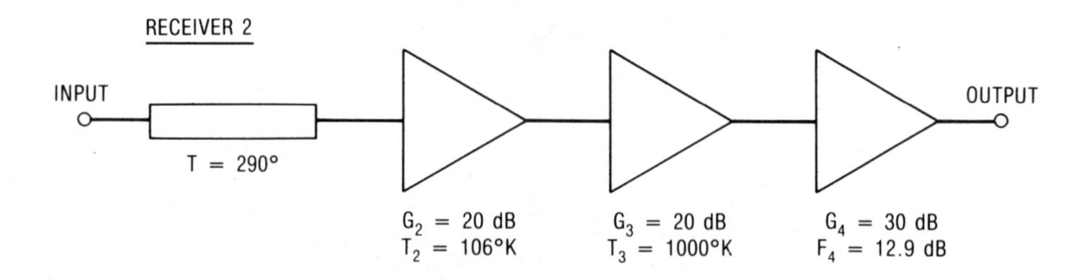

Assume the input noise in both cases is 4.5×10^{-14} W. The input S/N is 16.5 dB and the system bandwidth is 40 MHz.

3.15 Prove Parseval's theorem by direct substitution. Use the representation of the Dirac delta function of (3.22).

3.16 Show that the autocorrelation function of white noise with two-sided power spectral density $N_0/2$ is $N_0\delta(\tau)/2$.

3.17 The cross correlation function between realizations $x(t)$ and $y(t)$ of two different stationary random processes is: $r_{xy}(\tau) = E(x(t)y(t+\tau))$, and the cross spectral density is

$$W_{xy}(f) = \int_{-\infty}^{\infty} r_{xy}(\tau)e^{-i2\pi f\tau}\, d\tau$$

Show that $r_{yx}(\tau) = r_{xy}(-\tau)$ and $W_{yx}(f) = W_{xy}^{*}(f)$.

3.18 The integrals involving the factor k in (3.23) vanish because for $k \neq 0$ the exponential terms have many cycles over the frequency range B because $B\bar{\tau} \gg 1$, and positive and negative areas therefore cancel one another in the integration (stationary phase approximation). Show that this can also be established in the time domain by expressing $|H(f)|^2$ as a product of inverse Fourier tranforms and using (3.22) together with $B\bar{\tau} \gg 1$.

4

CONTINUOUS AND DISCRETE-TIME SIGNALS

4.1 THE SAMPLING THEOREM AND OVERSAMPLING

Almost all active sensing systems deal at some point in the processing path with discrete-time signals obtained by sampling continuous-time waveforms. For continuous-time signals occupying an effectively finite bandwidth the sampling theorem provides the discrete-time representation. As illustrated in Figure 4.1, let $x(t)$ be a band-limited function, which might represent a signal or noise, with Fourier transform

$$X(f_1) = \int_{-\infty}^{\infty} x(t) e^{-i2\pi f_1 t}\, dt \tag{4.1}$$

which is zero for $|f| > B$. If $x(t)$ is sampled at a rate $1/\Delta t$ a discrete-time series of the form $x(n\,\Delta t)$, $n = 0, 1, 2, \ldots$ is produced, where $x(n\,\Delta t)$ is given by

$$x(n\,\Delta t) = \int_{-\infty}^{\infty} X(f_1) e^{i2\pi f_1 n\,\Delta t}\, \mathrm{d}f_1 \tag{4.2}$$

The frequency-domain representation of the sampled discrete-time signal $x(n\,\Delta t)$ is given by the Fourier transform $X_{\mathrm{D}}(f)$ of $x(n\,\Delta t)$ which is defined as:

$$X_{\mathrm{D}}(f) = \sum_{n=-\infty}^{\infty} x(n\,\Delta t) e^{-i2\pi n f\,\Delta t} = \int_{-\infty}^{\infty} X(f_1) \sum_{n=-\infty}^{\infty} e^{-i2\pi n(f-f_1)\,\Delta t}\, df_1 \tag{4.3}$$

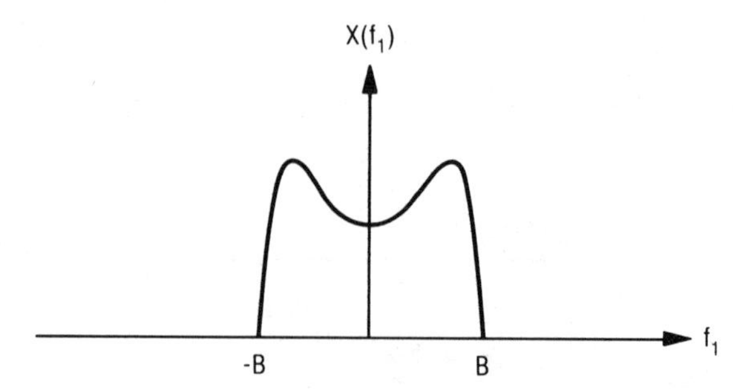

Figure 4.1 Spectrum of a band limited signal.

where we have used (4.2). And by making use of the identity (see Exercise 3.3)

$$\sum_{n=-\infty}^{\infty} e^{-i2\pi n(f-f_1)\,\Delta t} = \frac{1}{\Delta t} \sum_{n=-\infty}^{\infty} \delta\left(f - f_1 - \frac{n}{\Delta t}\right) \tag{4.4}$$

and substituting (4.4) into (4.3) yields the relationship between the frequency-domain representations of continuous and discrete-time signals:

$$X_{\mathrm{D}}(f) = \frac{1}{\Delta t} \sum_{n=-\infty}^{\infty} X\left(f - \frac{n}{\Delta t}\right) \tag{4.5}$$

where $X(f)$ is as defined in (4.1).

As illustrated in Figure 4.2, if $1/\Delta t > 2B$ there will be no overlap between the repeated spectra; overlap, which occurs if $1/\Delta t < 2B$, is referred to as *aliasing*. Therefore, if $1/\Delta t \geq 2B$ it should be possible to recover $x(t)$ from $X_{\mathrm{D}}(f)$ by band-limiting $X_{\mathrm{D}}(f)$ to $|f| \leq B$. In mathematical terms,

$$x(t) = \Delta t \int_{-B}^{B} X_{\mathrm{D}}(f) e^{i2\pi f t}\, df \tag{4.6}$$

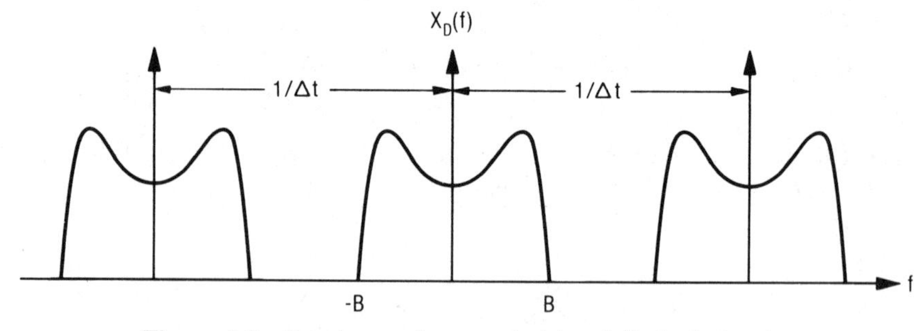

Figure 4.2 Spectrum of a sampled band limited signal.

And by substituting the first expression on the right-hand side of (4.3) into (4.6) $x(t)$ can be represented exactly as

$$x(t) = \Delta t \sum_{n=-\infty}^{\infty} x(n\,\Delta t)\left[\int_{-B}^{B} e^{+i2\pi f(t-n\,\Delta t)}\,df\right]$$

$$= \Delta t \sum_{n=-\infty}^{\infty} x(n\,\Delta t)\,\frac{\sin 2\pi B(t-n\,\Delta t)}{\pi(t-n\,\Delta t)} \tag{4.7}$$

As mentioned above, the minimum sampling rate for (4.7) to hold is the Nyquist rate $1/\Delta t = 2B$, and with this substitution (4.7) takes a more familiar form. However there are advantages to sampling at rates higher than the Nyquist rate, and also in using interpolation functions other than $\sin 2\pi Bt/\pi t$, both of which are sometimes mandatory.

Equation (4.7) demonstrates that a bandlimited continuous-time function contains considerable redundant information, since it can be completely characterized by and reconstructed from the samples $x(n\,\Delta t)$. For a signal of finite duration T we say that the signal can be approximately represented by $T/\Delta t$ samples, which is equal to $2BT$ for sampling at the Nyquist rate. This however is an approximation because strictly band-limited signals must be of infinite duration, which is reflected in the infinite limits in the summation in (4.7). More realistically, one can speak of a signal of finite duration T whose energy is sensibly contained (say ~90%) within some frequency range B. For example, the Fourier transform of a rectangular pulse extending over a duration $-T/2 \leq t \leq T/2$ is $\sin \pi fT/\pi f$, for which 90% of the signal energy is contained within the frequency range $|f| \leq 1/T$. However, the signal energy, however small, remains finite over the entire range $-\infty \leq f \leq \infty$ and therefore some degree of aliasing must occur. Hence there will always be some error in using the representation of (4.7), and the question arises concerning the accuracy with which (4.7) can be applied, and the necessary conditions under which application of (4.7) will be satisfactory.

As one of the conditions, it is generally necessary to sample at rates higher than the Nyquist rate $2B$.† Clearly, as is illustrated in Figure 4.3, the amount of unwanted signal energy in the tails of the neighboring repeated spectra which falls into the band of interest becomes reduced as the frequency separation between the repeated spectra increases, and, as seen in (4.5), this can be accomplished by increasing the sampling rate. In the limit as $\Delta t \rightarrow 0$ the sampled discrete-time signal approaches a continuous function and aliasing of course disappears.

One sometimes also employs anti-aliasing filtering for this purpose. This amounts to passing the signal through a low-pass filter in order to attenuate the spectral tails *before* sampling. Clearly, if the magnitude of the spectral

† The Nyquist rate in this case is somewhat arbitrary, being dependent on the criterion used for the definition of B. This is to be discussed shortly.

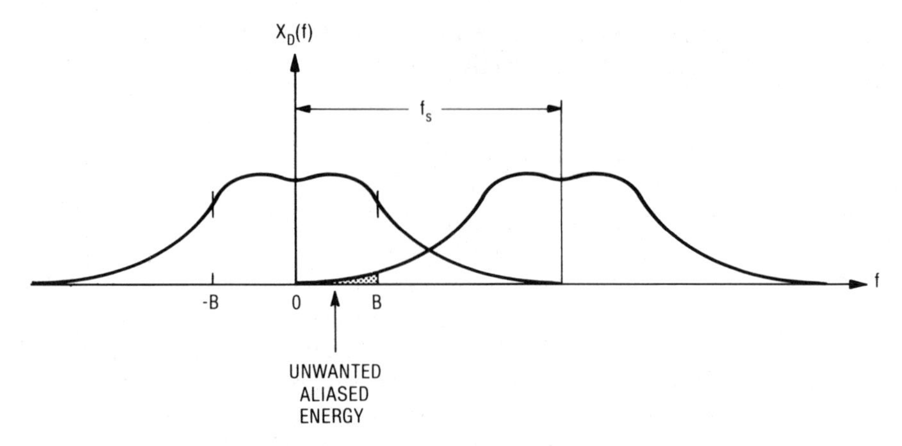

Figure 4.3 Illustration of aliasing.

tails is reduced, the distortion due to aliasing in the filtered signal will be less than the distortion due to aliasing in the unfiltered signal. This is of course possible only if the information in the spectral tails, which results in high-frequency fluctuations in the signal time function, can be dispensed with. As noted, anti-aliasing filtering must be applied before conversion of the continuous signal to a discrete-time signal. After sampling, the unwanted spectral tails in the band of interest cannot be reduced by filtering without also altering the spectrum of the signal itself. This is also clearly illustrated in Figure 4.3.

In addition to oversampling, the error can also be reduced by selecting an appropriate interpolation function. Equation (4.7) expresses the fact that the value of the continuous-time function $x(t)$ at the interpolation point t is equal to a two-sided weighted average of the samples $x(n\,\Delta t)$ on either side of t. As illustrated in Figure 4.4, the weights are equal to the values of the interpolation function at the sampling points when the interpolation function is positioned with its peak at t. For any finite sampling rate however, if the signal is of finite duration there can be only a finite number of points in the weighted average, which also contributes to the error. The interpolation function however always has the general form of $\sin \pi Bt/\pi t$; that is, there will be a main lobe with tails on either side which eventually become negligible. Therefore, as illustrated in Figure 4.5, the representation of (4.7) will be more accurate for values of t in the central portion of the signal, and will become less accurate as t moves towards the edges—the points $t = 0$ and $t = T$ for a signal of duration T—because the weighted average becomes increasingly one-sided and the number of points reduced. This is to be expected, since it is at the edges that the finite duration of the signal is most in evidence, whereas in the center the signal appears to the interpolation function to be more nearly infinite.

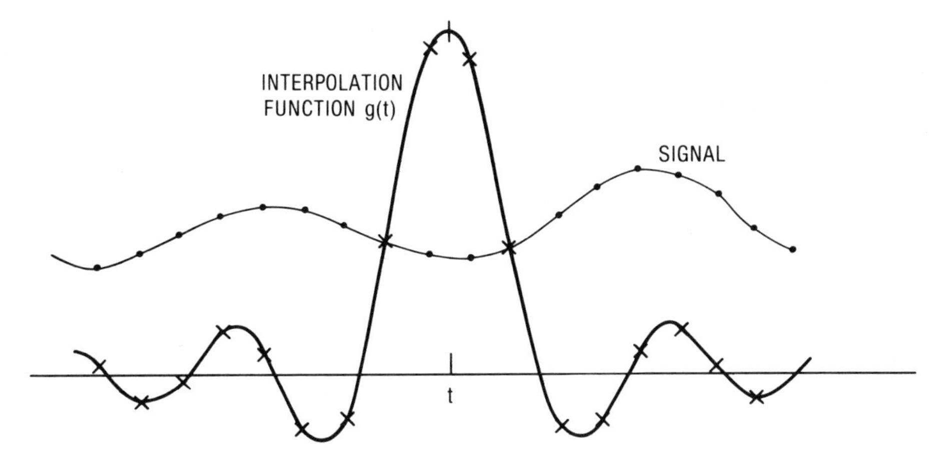

Figure 4.4 The interpolation process.

From this discussion it is clear that one wants an interpolation function whose area is concentrated as much as possible in the vicinity of its main lobe, and with tails that fall off as rapidly as possible, say, as $1/t^b$ where b is as large as possible. To see how this can be achieved let us rewrite (4.7) as

$$
\begin{aligned}
x(t) &= \sum_{n=n_1}^{n_2} x(n\,\Delta t) g(t - n\,\Delta t) \\
&= \sum_{n=n_1}^{n_2} x(n\,\Delta t) \int_{-\infty}^{\infty} G(f) e^{i2\pi f(t-n\,\Delta t)}\,dt
\end{aligned}
\tag{4.8}
$$

where $n_2 - n_1 + 1 = N$ is the number of points in the weighted average, being determined by the length of the interpolation function, and

$$
G(f) = \int_{-\infty}^{\infty} g(t) e^{-i2\pi ft}\,dt \tag{4.9}
$$

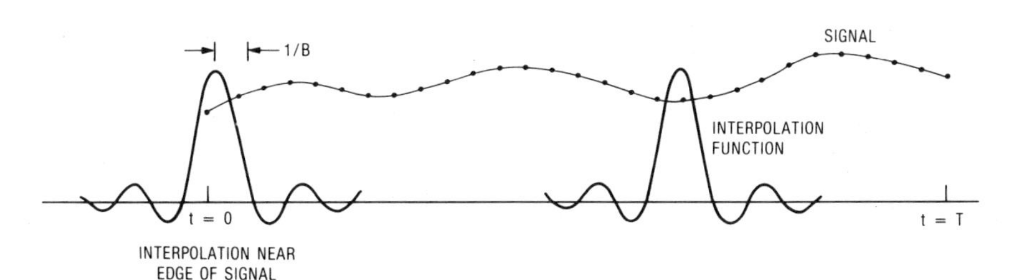

Figure 4.5 Illustration of edge effects.

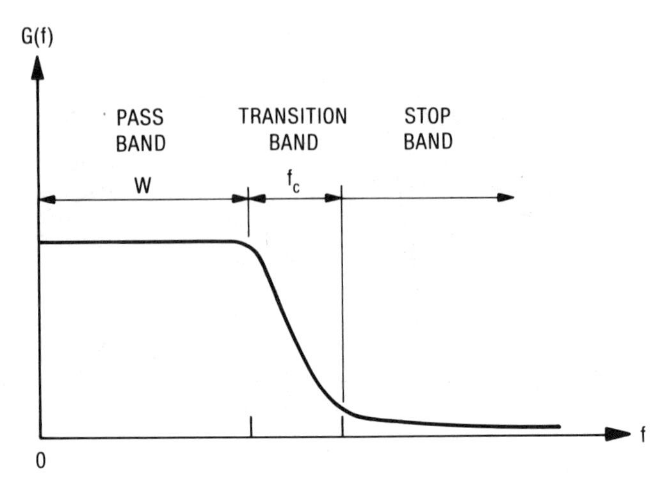

Figure 4.6 Critical filter bands.

The desired characteristics of the interpolation function $g(t)$ can be obtained by specifying $G(f)$, and we note in (4.8) that the interpolation process is exactly analogous to a filtering operation (see Section 4.5). Therefore, for this purpose, referring to Figure 4.6, we discuss some general considerations that arise in the design of filters.

As shown, there are essentially three regions of interest in $G(f)$: (1) the passband W, (2) the transition band f_c and (3) the stop band. For interpolation, $G(f)$ should be flat and equal to unity[†] over the passband $W \leq B$, and the magnitude of the stop band should be as small as possible to reject frequencies outside the band of interest, which in this case includes the repeated spectra in (4.5). Given any set of characteristics for (1) and (3), the nature of the transition band (2) will determine the number of terms in (4.8)—the value of N—which will be required to produce $G(f)$ to a sufficiently good approximation. Specifically, it can be shown that the longer and more gradual that the transition between the passband and the stop band can be made, the smaller will be the required value of N for the specified filter characteristic $G(f)$ to be realized satisfactorily. In addition, discontinuities in $G(f)$ should be avoided in order to enable as large a rate of fall-off as possible. To show this write $g(t) = \int_{-\infty}^{\infty} G(f)e^{i2\pi ft}\, df = \int_{-\infty}^{\infty} u\, dv$ with $u = G(f)$, $dv = e^{i2\pi ft} df$, and after integrating by parts

$$g(t) = G(f)\left.\frac{e^{i2\pi ft}}{i2\pi t}\right]_{-\infty}^{\infty} - \frac{1}{i2\pi t}\int_{-\infty}^{\infty} G'(f)e^{i2\pi ft}\, dt$$

$$= \frac{i}{2\pi t}\int_{-\infty}^{\infty} G'(f)e^{i2\pi ft}\, df$$

[†] Since the number of weights is finite, this requires renormalization so that their sum is unity.

because if $g(t)$ exists, then $\int_{-\infty}^{\infty} |G(f)|^2\, df < \infty$, which requires that $G(f) \to 0$ as $f \to \infty$. Also

$$|g(t)| = \frac{1}{|2\pi t|} \left| \int_{-\infty}^{\infty} G'(f) e^{i2\pi ft}\, dt \right| \le \frac{1}{|2\pi t|} \int_{-\infty}^{\infty} |G'(f)|\, df$$

Now, although $G'(f)$ does not exist at points of discontinuity, the integral $\int_{-\infty}^{\infty} G'(f)\, df$ does exist and can be evaluated since $G'(f)$ can be represented at discontinuities by delta functions. Hence, if $G(f)$ has discontinuities, $|g(t)|$ falls off at least as fast as $1/|t|$. For example, for $G(f)$ equal to unity for $|f| < B$ and zero otherwise, which has discontinuities at $f = \pm B$, $g(t) = \sin \pi Bt/\pi t$. However, faster rates of fall-off are possible and, in general, it can be shown (see Exercise 4.3) that if at least p derivatives of $G(f)$ exist for every point f (i.e., for some values of f the $(p+1)st$ derivative yields delta functions), then $g(t)$ falls off at least as fast as $1/t^{p+1}$.

The possibility of selecting desirable characteristics for $G(f)$ however also depends on the sampling rate, in a manner consistent with reducing effects of aliasing. This is illustrated in Figure 4.7a–b, where in (a) the sampling rate f_s is equal to the Nyquist rate $2B$, and in (b) f_s is greater than $2B$. In (a), $G(f)$ is equal to unity for $|f| \le B$ and is zero otherwise. Therefore, in fact, the transition region occupies a zero range of frequencies; there is no transition region. Also, dG/df yields delta functions so that $g(t)$ falls off as $1/t$, which is the slowest possible rate. Thus, sampling at the Nyquist rate yields the worst possible case in this sense. On the other hand, for $f_s > 2B$ it is seen in (b) that the transition region can be increased correspondingly which reduces the required value of N, and the discontinuities can also be avoided so that $g(t)$ will fall off faster than $1/t$. For example, if $G(f)$ employs a cosine roll off (see Exercise 4.4), $g(t)$ goes as $1/t^3$. Given a value of $f_s > 2B$, one must make a choice between the values of the passband W and transition range f_c under the nominal constraint $f_c + 2W = f_s$. That is, as W increases the main lobe of the interpolation function becomes narrower, but f_c must decrease, and so on. Specific values of f_c and W for any given value of f_s, and the resulting required value of N, will depend on the application, and can probably most easily be determined for any particular situation by experimenting at a computer terminal.

The interpolation errors at the edges of the signal can also be reduced by anti-aliasing filtering. The magnitude of the spectrum of the aforementioned rectangular pulse of duration T, $|\sin \pi fT/\pi f|$, falls off as $1/f$. On the other hand, if the pulse amplitude is, say, $(1 + \cos \pi t/T)$ for $|t| \le T/2$ and zero otherwise, its spectral tails go as $1/f^3$ (see Exercise 4.4). That is, as above, the smoother the variation in the pulse amplitude the smaller will be its high frequency content and therefore the smaller will be the spectral tails. Also, as is illustrated in Figure 4.8, the error associated with edge effects which accompanies the interpolation of signal values in the vicinity of a sharp abrupt signal edge will clearly be greater than in interpolating points over a smooth signal tail.

a) SAMPLING AT THE NYQUIST RATE

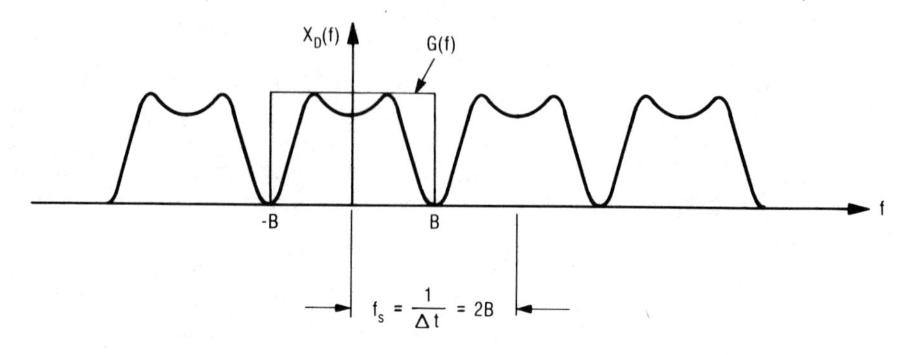

b) SAMPLING AT HIGHER THAN THE NYQUIST RATE

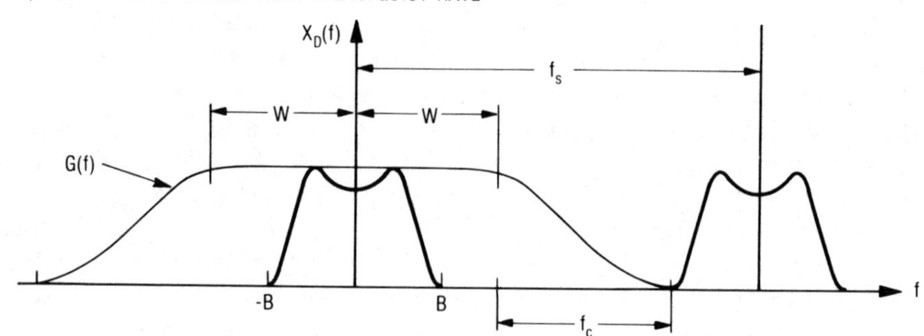

Figure 4.7 Illustration of effects of oversampling.

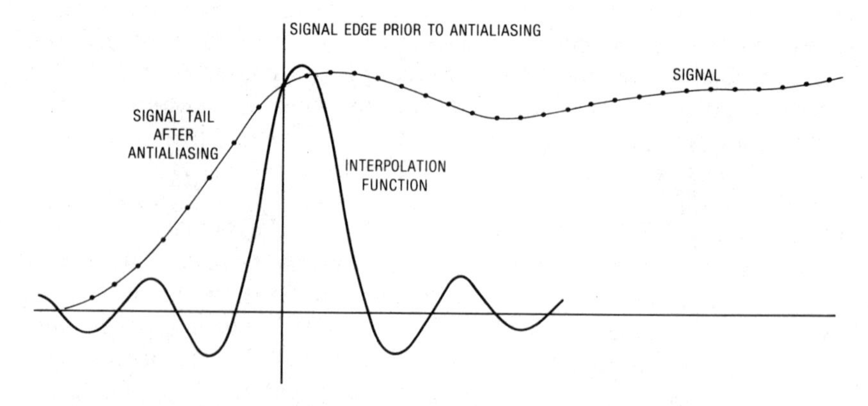

Figure 4.8 Amelioration of edge effects by antialiasing filter.

It is to be noted that in what follows we shall frequently assume sampling at the Nyquist rate and use the representation of (4.7) under the assumption of a band-limited signal. This is done for purposes of analysis and should not be viewed as a contradiction of the need for oversampling.

4.1.1 Application of the Sampling Theorem to Delay of Discrete-Time Signals

It is often necessary in the processing of sensor data to implement delays in discrete-time signals. This is of particular importance in digital beamforming of array antennas for radar, and also for sonar arrays. If the delays Δ are of the form $\Delta = I\,\Delta t$, where I is an integer this is of course very simple, since it involves only a reshuffling of sample time slots. On the other hand, it most often happens that $\Delta = (I + \eta)\,\Delta t$, where $0 < \eta < 1$, which presents a problem for a discrete-time series. The sampling theorem however provides a simple means for dealing with this.

Suppose, as illustrated in Figure 4.9a, a signal is to be delayed by a noninteger number of samples $I + \eta$; in this example $I = 4$ and $\eta = 1/3$. As

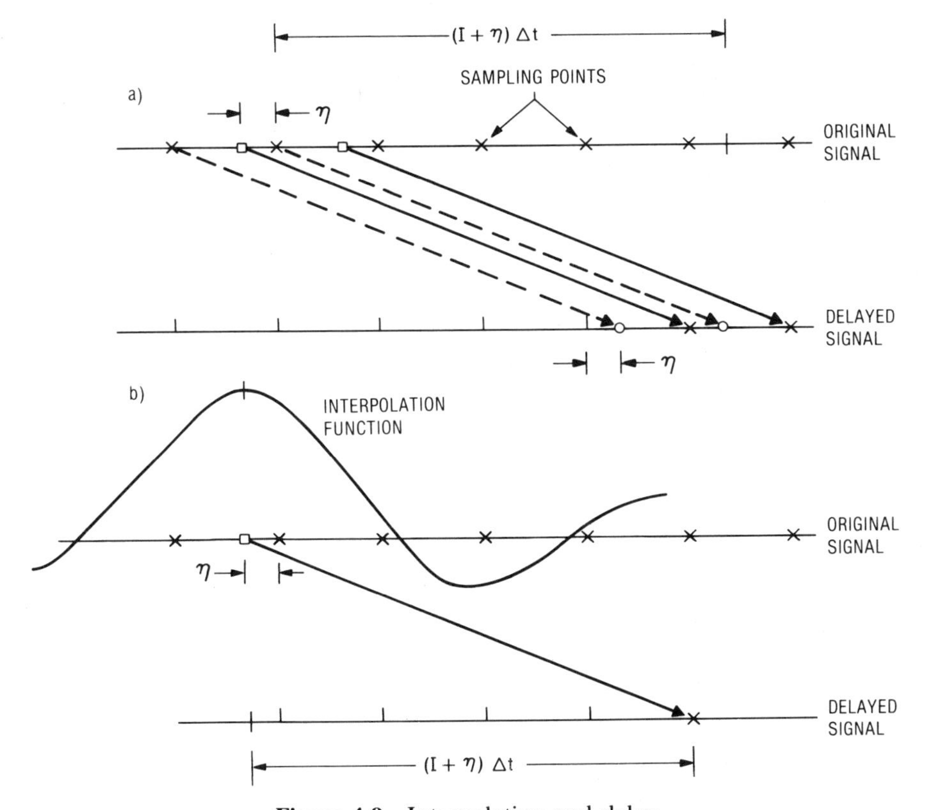

Figure 4.9 Interpolation and delay.

shown, the destination after delay of each original signal sample is a location in between sampling points, denoted by the 0s. But this is impossible since samples can only be at sampling points. This however is equivalent to placing at each sampling point the value the original signal had at the time $(I+\eta)\,\Delta t$ prior to the time corresponding to that sampling point, denoted by the open boxes (□). Of course there are no signal samples at these points, but, assuming $1/\Delta t$ is large enough, the value the original continuous signal had at the points □ prior to sampling can be reconstructed by using (4.8). Thus, as illustrated in Figure 4.9b, the procedure is to interpolate the original signal values at the □ by a weighted average of the samples on either side, and place the interpolated values in the appropriate delayed time slots. The value $S_{\mathrm{D}}(k)$ of the delayed signal at any time $k\,\Delta t$ is thus given by

$$S_{\mathrm{D}}(k) = \sum_{n=0}^{N-1} g_{\eta}(n) S\left(k - I - \frac{N}{2} + n\right) \tag{4.10}$$

where $S(k)$ is the signal value, and the $g_{\eta}(n)$ are the values of the interpolation function at the sample points after its peak has been positioned at the point to be interpolated. As noted above, there will be errors in the delayed signal, primarily at the edges, with the magnitudes of such errors depending on all the factors that have been discussed.

4.2 THE SAMPLING THEOREM FOR BANDPASS CARRIER SIGNALS

The foregoing results apply to video waveforms whose spectrum includes the origin, $f=0$. For bandpass waveforms a similar discussion and exactly the same principles apply, and, as before, a signal of duration T and bandwidth B can be approximately represented by a finite number of discrete samples if the sampling rate is high enough, in this case $1/\Delta t \geq B$. To show this we recall that a real bandpass signal $x(t)$ with carrier frequency f_0 can be written as

$$x(t) = \mathrm{Re}[h(t) e^{i2\pi f_0 t}] \tag{4.11}$$

with

$$X(f) = \int_{-\infty}^{\infty} x(t) e^{-i2\pi ft}\, dt = \frac{1}{2} H(f-f_0) + \frac{1}{2} H^*(-f-f_0) \tag{4.12}$$

and

$$H(f) = \int_{-\infty}^{\infty} h(t) e^{-i2\pi ft}\, dt$$

Since $x(t)$ is real, then $X(f) = X^*(-f)$, but $H(f)$ need not satisfy this relationship since $h(t)$ need not be real and, in particular, $|H(f)|$ need not be symmetrical about $f = 0$. Since the bandpass signal $x(t)$ has bandwidth B, then $H(f - f_0)$ and $H^*(-f - f_0)$ each vanish for $-B/2 > f - f_0 > B/2$ and $-B/2 > f + f_0 > B/2$ respectively.

Equations (4.2) and (4.3) then become

$$x(n\,\Delta t) = \mathrm{Re}[h(n\,\Delta t)e^{i2\pi f_0 n\,\Delta t}] \tag{4.13}$$

and

$$X_\mathrm{D}(f) = \sum_{n=-\infty}^{\infty} e^{-i2\pi n\,\Delta t}\,\mathrm{Re}[h(n\,\Delta t)e^{i2\pi f_0 n\,\Delta t}] = \frac{1}{2}\sum_{n=-\infty}^{\infty} h(n\,\Delta t)e^{-i2\pi(f-f_0)n\,\Delta t}$$
$$+ \frac{1}{2}\sum_{n=-\infty}^{\infty} h^*(n\,\Delta t)e^{-i2\pi(f+f_0)n\,\Delta t} = A(f) + B(f) \tag{4.14}$$

where $A(f)$ is the positive-frequency part of $X_\mathrm{D}(f)$ and $B(f)$ is the negative-frequency part. As discussed in connection with (4.5), for sampling at the Nyquist rate, in this case $1/\Delta t = B$, there will be no aliasing and, as before, $x(t)$ can be recovered by band-limiting $A(f)$ to $f_0 - B/2 \le f \le f_0 + B/2$ and $B(f)$ to $-f_0 - B/2 \le f \le -f_0 + B/2$, whence as in (4.7)

$$x(t) = \frac{\Delta t}{2}\sum_{n=-\infty}^{\infty} h(n\,\Delta t)\int_{f_0-B/2}^{f_0+B/2} e^{-i2\pi(f-f_0)n\,\Delta t}e^{i2\pi ft}\,df$$
$$+ \frac{\Delta t}{2}\sum_{n=-\infty}^{\infty} h^*(n\,\Delta t)\int_{-f_0-B/2}^{-f_0+B/2} e^{-i2\pi(f+f_0)n\,\Delta t}e^{i2\pi ft}\,df$$
$$= \frac{\Delta t}{2}\sum_{n=-\infty}^{\infty} h(n\,\Delta t)e^{i2\pi f_0 t}\,\frac{\sin \pi B(t - n\,\Delta t)}{\pi(t - \Delta t)}$$
$$+ \frac{\Delta t}{2}\sum_{n=-\infty}^{\infty} h^*(n\,\Delta t)e^{-i2\pi f_0 t}\,\frac{\sin \pi B(t - n\,\Delta t)}{\pi(t - n\,\Delta t)}$$
$$= \Delta t\sum_{n=-\infty}^{\infty} \mathrm{Re}(h(n\,\Delta t)e^{i2\pi f_0 t})\,\frac{\sin \pi B(t - n\,\Delta t)}{\pi(t - n\,\Delta t)} \tag{4.15}$$

This is of the same form as (4.7) but the samples $h(n\,\Delta t)$ are complex. If we write $h(t) = a(t)e^{i\phi(t)}$ (4.15) becomes

$$x(t) = \frac{\Delta t}{2}\sum_{n=-\infty}^{\infty} a(n\,\Delta t)\cos\phi(n\,\Delta t)\,\frac{\sin \pi B(t - n\,\Delta t)}{\pi(t - n\,\Delta t)}\cos 2\pi f_0 t$$
$$- \frac{\Delta t}{2}\sum_{n=-\infty}^{\infty} a(n\,\Delta t)\sin\phi(n\,\Delta t)\,\frac{\sin \pi B(t - n\,\Delta t)}{\pi(t - n\,\Delta t)}\sin 2\pi f_0 t \tag{4.16}$$

And, of course, samples of both the amplitude and the phase are required. Therefore, for a carrier signal of duration T, $T/\Delta t$ complex samples, or $2T/\Delta t$ real samples, are required to reconstruct the signal. Hence for sampling at the Nyquist rate $2BT$ real samples are required for carrier as well as video waveforms, although the Nyquist rate is different in the two cases for the same value of B. The discussion concerning the desirability of oversampling in order to enable more efficient and accurate interpolation holds equally in both cases.

4.3 SIGNAL DURATION AND BANDWIDTH

Since strictly bandlimited signals cannot exist, the bandwidth of signal is not a precisely defined, universally accepted, quantity. There are however a number of useful definitions, some of which are widely accepted. As noted above, we can speak of the frequency range within which most of the signal energy is contained. Another useful definition is in terms of the half-power or 3-dB width of the positive-frequency part of the Fourier transform.

Consider a video pulse of duration T. The magnitude squared of its Fourier transform is $|(\sin \pi fT)/\pi f|^2$, and its value at $f = 0$ is greater than its value at $f = 1/2T$ by a factor of $(\pi/2)^2 = 2.46$. Although the value at $f = 1/2T$ is therefore less than the value at the origin by 3.92 dB, one nevertheless speaks of the frequency range $0 \leq f \leq 1/2T$ as the 3-dB bandwidth of the video pulse. Thus, for a video signal of duration T we have the general rule of thumb that the 3-dB bandwidth is given by

$$B = \frac{1}{2T} \tag{4.17}$$

On the other hand, consider a sinusoid of frequency f_0, amplitude modulated by a rectangular pulse of duration $T \gg 1/f_0$. The magnitude-squared of the positive frequency part† of the signal spectrum is to a very good approximation $|\sin \pi T(f - f_0)]/\pi(f - f_0)|^2$. Hence, by following the same line of argument applied to video pulses, the 3-dB signal bandwidth in this case is

$$B = \frac{1}{T} \tag{4.18}$$

Thus, excluding the large time-bandwidth signals discussed in Chapter 9, (4.17) and (4.18) express generally accepted definitions for the bandwidth of video and carrier pulses, for which BT is of order unity.

In addition to the bandwidth, the duration of a signal can also be subject to interpretation, since in practice, pulses are never strictly rectangular and

† Ignoring the tails of the negative-frequency component $[\sin \pi T(f + f_0)]/\pi(f + f_0)$.

can in fact have tails of significant duration. Another useful set of definitions, for the duration Δt and bandwidth Δf of signal $x(t)$, are

$$\Delta f = \left(\frac{4\pi^2 \int_{-\infty}^{\infty} f^2 |X(f)|^2 \, df}{\int_{-\infty}^{\infty} |X(f)|^2 \, df} \right)^{1/2}$$

$$\Delta t = \left(\frac{\int_{-\infty}^{\infty} t^2 |x(t)|^2 \, dt}{\int_{-\infty}^{\infty} |x(t)|^2 \, dt} \right)^{1/2} \tag{4.19}$$

where $X(f) = \int_{-\infty}^{\infty} x(t) e^{-i2\pi ft} \, dt$.

The relationship between Δf and Δt is as follows. By the Schwarz inequality

$$\left| \int_{-\infty}^{\infty} tx(t)x'(t)dt^2 \le \int_{-\infty}^{\infty} t^2 |x(t)|^2 \, dt \int_{-\infty}^{\infty} |x'(t)|^2 \, dt \right. \tag{4.20}$$

and

$$x'(t) = -i2\pi \int_{-\infty}^{\infty} fX(f) e^{-i2\pi ft} \, df$$

Hence, by using (see Section 4.5)

$$\int_{-\infty}^{\infty} e^{i2\pi ft} \, dt = \delta(f)$$

it follows by direct substitution that

$$\int_{-\infty}^{\infty} |x'(t)|^2 \, dt = 4\pi^2 \int_{-\infty}^{\infty} f^2 |X(f)|^2 \, df \tag{4.21}$$

By integrating by parts

$$\int_{-\infty}^{\infty} tx(t)x'(t) \, dt) = \frac{tx^2(t)}{2} \Big]_{-\infty}^{\infty} - \frac{1}{2} \int_{-\infty}^{\infty} x^2(t) \, dt$$

and

$$\left| \int_{-\infty}^{\infty} tx(t)x'(t) \, dt \right|^2 = \frac{1}{4} \left(\int_{-\infty}^{\infty} x^2(t) \, dt \right)^2 \tag{4.22}$$

because $x^2(t)$ goes to zero for large t faster than $1/t$ for the signals of interest (see Section 4.1). Therefore, since by Parseval's theorem (3.79): $\int_{-\infty}^{\infty} |x(t)|^2 = \int_{-\infty}^{\infty} |X(f)|^2 \, dt$, we obtain, by combining (4.19), (4.20), (4.21) and (4.22)

$$\Delta f \, \Delta t \ge \frac{1}{2} \tag{4.23}$$

with equality if $tx(t) \propto x'(t)$, in which case $x(t) \propto e^{-t^2/2}$, a Gaussian pulse.

Equation (4.23) is not inconsistent with $BT \approx 1$, and states that although this is a lower bound on the time-bandwidth product, there is no upper bound. We explore the properties and possible advantages of large time–bandwidth waveforms in Chapter 9.

4.4 THE ANALYTIC SIGNAL

The analytic signal is a complex representation of a real waveform that can facilitate certain types of computations which arise frequently in many branches of applied mathematics. Consider a function $f(t)$ with Fourier transform $F(f)$. The analytic signal $z(t)$ corresponding to $f(t)$ can be defined in terms of its Fourier transform $Z(f)$ as

$$Z(f) = 2F(f)V(f) \tag{4.24}$$

where $V(f)$ is here the unit frequency step function.

$$V(f) = \begin{cases} 1 & f \geq 0 \\ 0 & \text{otherwise} \end{cases}$$

Thus $Z(f)$ is equal to zero for $f < 0$. Also (see Section 4.5)

$$z(t) = 2\int_{-\infty}^{\infty} f(\tau)v(t-\tau)\, d\tau$$

where the inverse Fourier transform $v(t)$ of $V(f)$ is

$$v(t) = \int_{0}^{\infty} e^{i2\pi ft}\, df \tag{4.25}$$

It will now be shown that

$$v(t) = \frac{i}{2\pi t} + \frac{\delta(t)}{2} \tag{4.26}$$

In order to do this we make use of the Riemann–Lebesgue lemma (see Exercise 4.7) which states that if $q(t)$ is continuous and its first derivative exists over any interval (a, b) then

$$\lim_{f\to\infty} \int_{a}^{b} q(t)e^{-i2\pi ft}\, dt \to 0 \tag{4.27}$$

Because of (4.27) the function $\lim_{f\to\infty} (\pm\, e^{\pm i2\pi ft}/i2\pi t)$ is equivalent to $\delta(t)/2$. To show this, note that if $g(t)$ is continuous and $dg(t)/dt$ exists then for any value of ϵ

$$\int_{-\infty}^{\infty} \lim_{f\to\infty} \frac{e^{i2\pi ft}}{i2\pi t}\, dt = \lim_{f\to\infty} \int_{-\infty}^{-\epsilon} \frac{g(t)}{i2\pi t}\, e^{i2\pi ft}\, dt + \lim_{f\to\infty} \int_{+\epsilon}^{\infty} \frac{g(t)}{i2\pi t}\, e^{i2\pi ft}\, dt$$
$$+ \lim_{f\to\infty} \int_{-\epsilon}^{\epsilon} g(t)\, \frac{e^{i2\pi ft}}{i2\pi t}\, dt = \lim_{f\to\infty} \int_{-\epsilon}^{\epsilon} g(t)\, \frac{e^{i2\pi ft}}{i2\pi t}\, dt \tag{4.28}$$

because the first two terms on the right-hand side of (4.28) vanish by (4.27) since $g(t)/t$ is continuous and has a first derivative everywhere except at $t = 0$. Now let ϵ become arbitrarily small, while remaining finite. Then

$$\lim_{f\to\infty} \int_{-\epsilon}^{\epsilon} g(t)\, \frac{e^{i2\pi ft}}{i2\pi t}\, dt \to \lim_{f\to\infty} g(0) \int_{-\epsilon}^{\epsilon} \frac{\cos 2\pi ft}{i2\pi t}\, dt$$
$$+ \lim_{f\to\infty} g(0) \int_{-\epsilon}^{\epsilon} \frac{\sin 2\pi ft}{2\pi t}\, dt \tag{4.29}$$

because, since $g(t)$ is continuous it can be made arbitrarily close to $g(0)$ for ϵ sufficiently small.

The first term on the right-hand side of (4.29) vanishes since $\cos 2\pi ft/2\pi t$ is an odd function, and by making the substitution $x = 2\pi ft$ the second term becomes

$$\lim_{f\to\infty} g(0) \int_{-\epsilon}^{\epsilon} \frac{\sin 2\pi ft}{2\pi t}\, dt = g(0) \lim_{f\to\infty} \frac{1}{2\pi} \int_{-2\pi\epsilon f}^{2\pi\epsilon f} \frac{\sin x}{x}\, dx$$
$$= \frac{g(0)}{2\pi} \int_{-\infty}^{\infty} \frac{\sin x}{x}\, dx = \frac{g(0)}{2} \tag{4.30}$$

because $\int_{-\infty}^{\infty} (\sin x/x)\, dx = \pi$.

This is the desired result, because from (4.28), (4.29) and (4.30)

$$\int_{-\infty}^{\infty} g(t) \lim_{f\to\infty} \frac{e^{i2\pi ft}}{i2\pi t}\, dt = \frac{g(0)}{2} = \int_{-\infty}^{\infty} g(t)\, \frac{\delta(t)}{2}\, dt \tag{4.31}$$

Now returning to (4.25)

$$v(t) = \int_{-\infty}^{\infty} V(f) e^{i2\pi ft}\, df = \lim_{f\to\infty} \int_{0}^{f} e^{i2\pi xt}\, dx = \lim_{f\to\infty} \frac{e^{i2\pi ft} - 1}{i2\pi t}$$
$$= \frac{i}{2\pi t} + \frac{\delta(t)}{2} \tag{4.32}$$

and therefore

$$z(t) = 2 \int_{-\infty}^{\infty} f(\tau) v(t-\tau)\, d\tau = f(t) + \frac{i}{\pi} \int_{-\infty}^{\infty} \frac{f(\tau)}{(t-\tau)}\, d\tau = f(t) + i\hat{f}(t) \tag{4.33}$$

where $\hat{f}(t)$ denotes the Hilbert transform of $f(t)$.

$$\hat{f}(t) = \frac{1}{\pi}\int_{-\infty}^{\infty} \frac{f(\tau)}{t-\tau}\, d\tau \tag{4.34}$$

Hence $f(t)$ is the real part of the analytic signal $z(t)$ and $\hat{f}(t)$ is the imaginary part of $z(t)$.

In determining the Hilbert transform of a signal it is often easier to work in the frequency domain. Equation (4.24) can be written as

$$Z(f) = F(f) + \mathrm{sgn}(f)F(f) \tag{4.35}$$

where

$$\mathrm{sgn}(f) = \begin{cases} +1 & f \geq 0 \\ -1 & f < 0 \end{cases}$$

So, for example, if $f(t) = \cos 2\pi f_0 t$ then the Fourier transform of $\hat{f}(t)$ is $(\delta(f - f_0) - \delta(f + f_0))/i2$ and $\hat{f}(t) = \sin 2\pi f_0 t$.

The analytic signal can facilitate calculations such as

$$f(t)\cos 2\pi f_0 t * g(t)\sin 2\pi f_0 t = \int_{-\infty}^{\infty} f(\tau)g(t-\tau)\cos 2\pi f_0\tau \sin 2\pi f_0(t-\tau)\, d\tau \tag{4.36}$$

which would otherwise be quite cumbersome. To show this let $c(t) = a(t) * b(t)$ with

$$Z_c(f) = 2C(f)V(f)$$
$$Z_a(f) = 2A(f)V(f)$$
$$Z_b(f) = 2B(f)V(f)$$

and $A(f)$, $B(f)$ and $C(f)$ are the Fourier transforms of $a(t)$, $b(t)$ and $c(t)$ which are real functions of time, and $Z_a(f)$, $Z_b(f)$ and $Z_c(f)$ are the Fourier transforms of $z_a(t)$, $z_b(t)$ and $z_c(t)$, which are the analytic signals corresponding to $a(t)$, $b(t)$ and $c(t)$, respectively. Then since $C(f) = A(f)B(f)$, $Z_c(f)$ is

$$\begin{aligned} Z_c(f) &= 2C(f)V(f) = 2A(f)B(f)V(f) = A(f)Z_b(f) = B(f)Z_a(f) \\ &= \tfrac{1}{2}Z_a(f)Z_b(f) \end{aligned}$$

because both $Z_b(f)$ and $Z_a(f)$ vanish for $f < 0$. Then

$$z_c(t) = a(t) * z_b(t) = b(t) * z_a(t) = \tfrac{1}{2}z_a(t) * z_b(t) \tag{4.37}$$

But from (4.33) $c(t) = \text{Re}[z_c(t)]$, which leads to the very useful result

$$c(t) = a(t) * b(t) = \text{Re}(z_c(t)) = \text{Re}(a(t) * z_b(t)) = \text{Re}(b(t) * z_a(t))$$
$$= \tfrac{1}{2}\,\text{Re}(z_a(t) * z_b(t)) \tag{4.38}$$

Using analytic signals, Parseval's theorem is

$$\int_{-\infty}^{\infty} a^2(t)\,dt = \int_{-\infty}^{\infty} \hat{a}^2(t)\,dt = E = \frac{1}{2}\int_{-\infty}^{\infty} |z_a(t)|^2\,dt = \frac{1}{2}\int_{-\infty}^{\infty} |Z_a(f)|^2\,df \tag{4.39}$$

where E is the signal energy.

Another very useful result (see Exercise 4.9) is that if $a(t)$ is band-limited to $|f| < B/2$ with $B < 2f_0$, then,

$$HTa(t)\cos 2\pi f_0 = a(t)HT\cos 2\pi f_0 t = a(t)\sin 2\pi \mathrm{f}_0 \mathrm{t} \tag{4.40}$$

with of course the same result holding for $a(t)\sin 2\pi f_0 t$; here HT denotes the Hilbert transform operation, (4.34). This relationship is used extensively in Chapter 7.

4.5 PROCESSING OF CONTINUOUS AND DISCRETE-TIME SIGNALS

A continuous-time, linear time invariant system is described by an impulse response $h(t)$, which is the output of the system when the input is an impulse $\delta(t)$ applied at $t = 0$. The function $\delta(t)$ is the Dirac delta function which has the properties

$$\delta(t) = 0 \quad \text{for } t \neq 0$$
$$\int_{-\infty}^{\infty} \delta(t)\,dt = 1 \tag{4.41}$$
$$\int_{-\infty}^{\infty} f(t)\delta(t-a)\,dt = f(a)$$

At $t = 0$, $\delta(t)$ becomes infinite. The delta function can be represented as

$$\delta(t) = \int_{-\infty}^{\infty} e^{i2\pi ft}\,dt \tag{4.42}$$

because if $f(t)$ and $F(f)$ are Fourier-transform pairs then

$$f(t) = \int_{-\infty}^{\infty} F(f)e^{i2\pi ft}\,df = \int_{-\infty}^{\infty} f(t_1)\left[\int_{-\infty}^{\infty} e^{-i2\pi f(t_1 - t)}\,df\right]dt_1$$

from which (4.42) follows by (4.41).

Because the system is time-invariant, the response to an impulse $\delta(t-\tau)$ applied at $t=\tau$ is $h(t-\tau)$, simply a time-delayed version of $h(t)$. If the system response varied with time, which we do not deal with here, the response would be written as $h(t, \tau)$.

Consider an input $x(t)$ consisting of a train of impulses of strength $x(n\,\Delta t)$

$$x(t) = \sum_{n=-\infty}^{\infty} x(n\,\Delta t)\delta(t - n\,\Delta t)\,\Delta t \tag{4.43}$$

Since $\delta(t)$ has dimensions $1/t$ the factor Δt makes (4.43) dimensionally correct and also avoids infinities in $x(t)$ as $\Delta t \to 0$. Now each $\delta(t - n\,\Delta t)$ when applied to the system yields $h(t - n\,\Delta t)$ by definition of $h(t)$, hence

$$\begin{aligned} y(t) &= \lim_{\Delta t \to 0} \sum_{n=-\infty}^{\infty} x(n\,\Delta t)h(t - n\,\Delta t)\,\Delta t \\ &= \int_{-\infty}^{\infty} x(\tau)h(t-\tau)\,d\tau = \int_{-\infty}^{\infty} x(t-\tau)h(\tau)\,d\tau \\ &= x(t) * h(t) \end{aligned} \tag{4.44}$$

which is the convolution of $x(t)$ with $h(t)$, generally denoted as $*$.

If the system is causal there can be no output before the input arrives, so that $h(t)=0$ for $t<0$, or $h(t-\tau)=0$ for $t<\tau$, and (4.44) becomes

$$y(t) = \int_{-\infty}^{t} x(t)h(t-\tau)\,d\tau = \int_{0}^{\infty} x(t-\tau)h(\tau)\,d\tau$$

If, in addition, $x(t)=0$ for $t<0$, then

$$y(t) = \int_{0}^{t} x(t)h(t-\tau)\,d\tau = \int_{0}^{t} x(t-\tau)h(\tau)\,d\tau \tag{4.45}$$

which of course can also be written as in (4.44) if these conditions on $x(t)$ and $h(t)$ are understood.

The impulse response is equivalent to a Green's function for a linear differential operator, L, which is an equivalent way to describe the system.

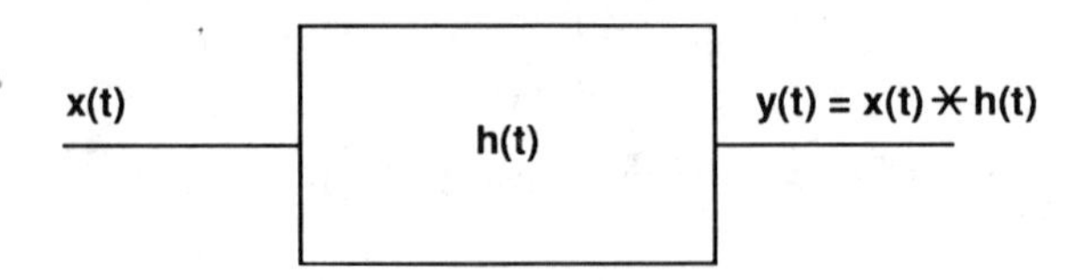

Figure 4.10 Continuous-time filtering.

If

$$L[y(t)] = x(t) \tag{4.46}$$

with $y(0) = y(1) = 0$, then $y(t)$ is given by

$$y(t) = \int_0^1 G(t, \tau)x(\tau)\, d\tau \tag{4.47}$$

where $G(t, \tau)$ is the solution of

$$L[G(t, \tau)] = \delta(t - \tau) \tag{4.48}$$

That is, applying $L[\ ]$ to (4.47) yields

$$L[y(t)] = \int_0^1 L[G(t, \tau)]x(\tau)\, d\tau = \int_0^1 \delta(t - \tau)x(\tau)\, d\tau = x(t) \tag{4.49}$$

Equation (4.47) applies generally to differential operators L with fixed or time-varying coefficients. However, if the coefficients are constant then $G(t, \tau) = G(t - \tau)$.

Filter characteristics are usually specified in terms of their transfer function $H(f)$, which is the Fourier transform of $h(t)$. Taking the Fourier transform of (4.44) yields the convolution theorem

$$\int_{-\infty}^{\infty} h(t) * x(t)e^{-i2\pi ft}\, dt = Y(f) = H(f)X(f) \tag{4.50}$$

where $X(f)$ and $Y(f)$ are the Fourier transforms of $x(t)$ and $y(t)$. Thus the Fourier transform of the output is simply the Fourier transform of the input multiplied by the transfer function.

In discrete-time systems we deal with sampled quantities $x(n\,\Delta t)$, $y(n\,\Delta t)$ and $h(n\,\Delta t)$ where n is an integer. Dropping the Δt for convenience, the equations corresponding to (4.44) and (4.45) for output $y(n)$ is

$$y(n) = \sum_{k=0}^{n} x(k)h(n - k) = \sum_{k=0}^{n} h(k)x(n - k) = \sum_{k=-\infty}^{\infty} h(k)x(n - k) \tag{4.51}$$

where $h(n)$ is the response of the system to a unit impulse

$$\delta_{n,0} = \begin{cases} 1, & n = 0 \\ 0, & n \neq 0 \end{cases}$$

If $h(t)$ and $x(t)$ in (4.45) are of duration T_1 and T_2, then $y(t)$ will be of duration $T_1 + T_2$. Similarly, as is illustrated in Figure 4.11, if $h(n)$ and $x(n)$

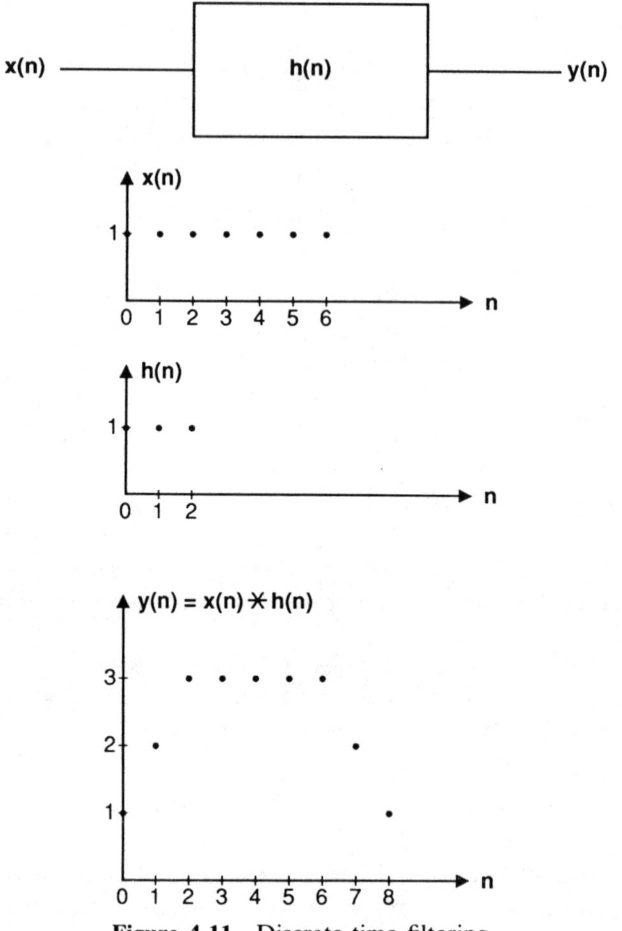

Figure 4.11 Discrete-time filtering.

are of duration M and N respectively, the output $y(n)$ is of duration $M + N - 1$. Equation (4.50) also holds for the Fourier transforms $Y(f)$, $H(f)$ and $X(f)$ of discrete-time signals $y(n)$, $h(n)$, $x(n)$ defined in (4.3).

In sampled-data discrete-time systems the Fourier transforms are also discretized and we write, referring to (4.3)

$$X(m\,\Delta f) = \sum_{n=0}^{\infty} x(n)e^{-i2\pi n\,\Delta t m\,\Delta f} \tag{4.52}$$

If $x(t)$ is of duration $T = N\,\Delta t$ the frequency resolution Δf is $1/T$ (see Exercise 4.13), whence $\Delta t\,\Delta f = 1/N$ and (4.52) can be written as

$$X(m) = \sum_{n=0}^{N-1} x(n) \exp\left(-i2\pi \frac{nm}{N}\right) = X(m + N) \tag{4.53}$$

As before (4.5), the discretized Fourier transform, in which we drop the subscript D, is periodic. The period in this case is $N = T/\Delta t$ because, referring to (4.5),

$$X\left(f - \frac{1}{\Delta t}\right) \rightarrow X\left(m\,\Delta f - \frac{1}{\Delta t}\right) = X\left(\frac{1}{T}\left(m - \frac{T}{\Delta t}\right)\right) \rightarrow X(m - N)$$

Equation (4.53) defines the Discrete Fourier Transform (DFT) of $x(n)$. The inverse DFT, $(\mathrm{DFT})^{-1}$, is defined as

$$\hat{x}(n) = \frac{1}{N} \sum_{m=0}^{N-1} X(m) \exp\left(i2\pi \frac{nm}{N}\right) \tag{4.54}$$

and by substituting (4.53) into (4.54)

$$\begin{aligned} \hat{x}(n) &= \frac{1}{N} \sum_{m=0}^{N-1} \sum_{k=0}^{N-1} x(k) \exp\left(-i2\pi \frac{km}{N}\right) \exp\left(i2\pi \frac{mn}{N}\right) \\ &= \frac{1}{N} \sum_{k=0}^{N-1} x(k) \sum_{m=0}^{N-1} \exp\left[-i2\pi \frac{m}{N}(n-k)\right] \end{aligned}$$

By using

$$\sum_{m=0}^{N-1} r^m = \frac{1 - r^N}{1 - r}$$

we obtain

$$\sum_{m=0}^{N-1} \exp\left[-i2\pi \frac{m}{N}(n-k)\right] = \exp\left[-i\pi\left(\frac{N-1}{N}\right)(n-k)\right] \frac{\sin \pi(n-k)}{\sin \frac{\pi}{N}(n-k)} \tag{4.55}$$

which is equal to N if $n - k = pN$, $p = \ldots, -2, -1, 0, 1, 2, \ldots$, and is zero otherwise. Hence

$$\hat{x}(n) = \sum_{l=-\infty}^{\infty} x(n - pN) \tag{4.56}$$

Thus, applying DFT and $(\mathrm{DFT})^{-1}$ successively to $x(n)$ produces a periodic version of the original waveform, with period N. Of course $\hat{x}(n) = x(n)$ for $n = 0, 1, 2, \ldots, N-1$.

In continuous-time systems, (4.44) describes what is implemented in practice. In discrete-time system the corresponding equation (4.51) is sometimes implemented in practice but often is not. In an alternative approach, the speed of the Fast Fourier Transform (FFT) algorithm enables

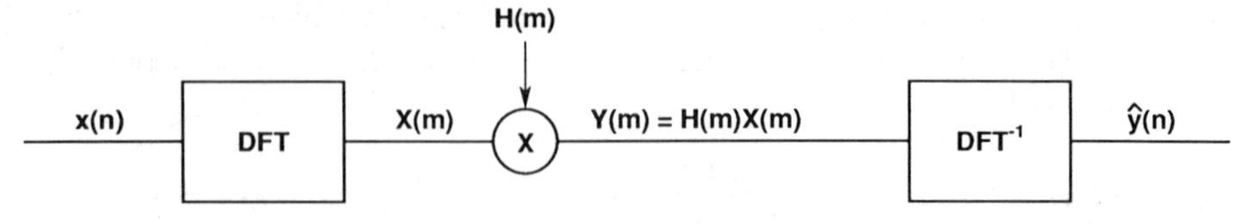

Figure 4.12 FFT filtering.

filtering operations to be carried out much more rapidly than the time-domain operation of (4.51). The procedure is illustrated in Figure 4.12. We note however, referring to (4.54) and (4.56), that

$$\hat{y}(n) = \frac{1}{L}\sum_{m=0}^{L-1} X(m)H(m)\exp\left(i2\pi\frac{mn}{L}\right) = \frac{1}{L}\sum_{m=0}^{L-1} Y(m)\exp\left(i2\frac{\pi mn}{L}\right) \tag{4.57}$$

is periodic, with period L, and since $y(n)$ is of duration $M + N - 1$, then unless $L \geq M + N - 1$ the resulting $\hat{y}(n)$ will be aliased in time, in exactly the same way that $X_D(f)$ in (4.5) will be aliased in frequency if $1/\Delta t$ is not large enough. This requires that $X(m)$ and $H(m)$ be implemented as $X'(m)$ and $H'(m)$, given by

$$\begin{aligned} X'(m) &= \sum_{n=0}^{L-1} x(n)\exp\left(-i2\pi\frac{nm}{L}\right) \\ H'(m) &= \sum_{n=0}^{L-1} H(n)\exp\left(-i2\pi\frac{nm}{L}\right) \end{aligned} \tag{4.58}$$

where $L \geq M + N - 1$.

Of course $x(n)$ and $h(n)$ are non-zero only over the ranges $0 \leq n \leq N-1$ and $0 \leq n \leq M-1$, both of which are less than L. The effect of including the additional zero points in the DFT is as follows. We can write

$$\begin{aligned} X'(m) &= \sum_{n=0}^{N-1} x(n)\exp\left[-i2\pi\frac{n}{N}\left(m\frac{N}{L}\right)\right] \\ &= \sum_{n=0}^{N-1} x(n)\exp\left(-i2\pi\frac{n}{N}m'\right) = X(m') \end{aligned} \tag{4.59}$$

where $m' = Nm/L$. Thus, whereas $X(m)$ contains frequency information at N integer values of m, the number of points in $X'(m)$ is greater by a factor of L/N. For example, if $L = 2N$ then $X'(m) = X(m/2)$ has an additional value in between each of the original frequency points of $X(m)$. This however does not represent any new information. That is, continuing with this example, using (4.54) and (4.55), (4.59) can be written as

$$
\begin{aligned}
X'(m) &= \sum_{n=0}^{N-1} \frac{1}{N} \sum_{k=0}^{N-1} X(k) \exp\left(i2\pi \frac{kn}{N}\right) \exp\left(-i2\pi \frac{nm}{L}\right) \\
&= \frac{1}{N} \sum_{k=0}^{N-1} X(k) \exp\left[i\pi\left(\frac{N-1}{N}\right)(k - m/2)\right] \frac{\sin \pi(k - m/2)}{\sin \frac{\pi}{N} (k - m/2)}
\end{aligned}
\tag{4.60}
$$

and $X'(m)$ is identical with $X(k)$ at the original points represented by $m = 2k$. For example, if $m = 4$, then $[\sin \pi(k-2)]/[\sin(\pi/N)(k-2)]$ is equal to zero unless $k = 2$ and $X'(4) = X(2)$. The additional points in between, in this case for $m = 1, 3, 5, \ldots$, are interpolated values consisting of a weighted average of values of $X(k)$. The weights are equal to the values of the interpolation function

$$
\exp\left[i\pi\left(\frac{N-1}{N}\right)(k - m/2)\right] \frac{\sin \pi(k - m/2)}{\sin \frac{\pi}{N} (k - m/2)}
$$

at the original frequency points of $X(k)$ when the peak of the interpolation function is positioned at $m/2$, which of course need not be an integer. The value of $X'(m)$ for even values of m can also be viewed as an interpolation, with all weights equal to zero except that for which $k = 2m$. Thus, $\hat{y}(n)$ in (4.57), with $H(m)$ and $X(m)$ implemented as in (4.58), is identical to $y(n)$ of (4.51) for $0 \leq n \leq M + N - 1$.

With FFT filtering of very long signals such as speech waveforms, another problem arises which does not occur with convolutional time domain-filtering (4.51). Obviously, FFTs cannot be of infinite duration and as shown in Fig. (4.13) it is therefore necessary to break up the input into relatively short segments, say of length N, which are processed individually. However, if the three segments shown were dealt with as a single unit, in the process of convolution, points in the vicinity of the leading edge of the middle segment would contain contributions from points in the vicinity of the trailing edge of the first segment. This of course cannot take place when the segments are processed individually. However, as illustrated in Figure 4.13, it is possible to achieve the desired effect by appropriately fitting together the individually processed segments, such that the last $M - 1$ points of each processed segment overlap the first $M - 1$ points of the following processed segment, and summing the values of the overlapping points. In this way, the three segments each of length $M + N - 1$ which are processed separately can be joined to produce a single segment of length $3N + M - 1$ which is identical to that which would have been produced if a single segment of length $3N$ had been processed as a single unit. In this manner sequences of effectively infinite duration can be processed using FFTs. The foregoing procedure is known as the Overlap Add method. An alternative approach known as Overlap Save, which we do not discuss here, yields the same net result [17].

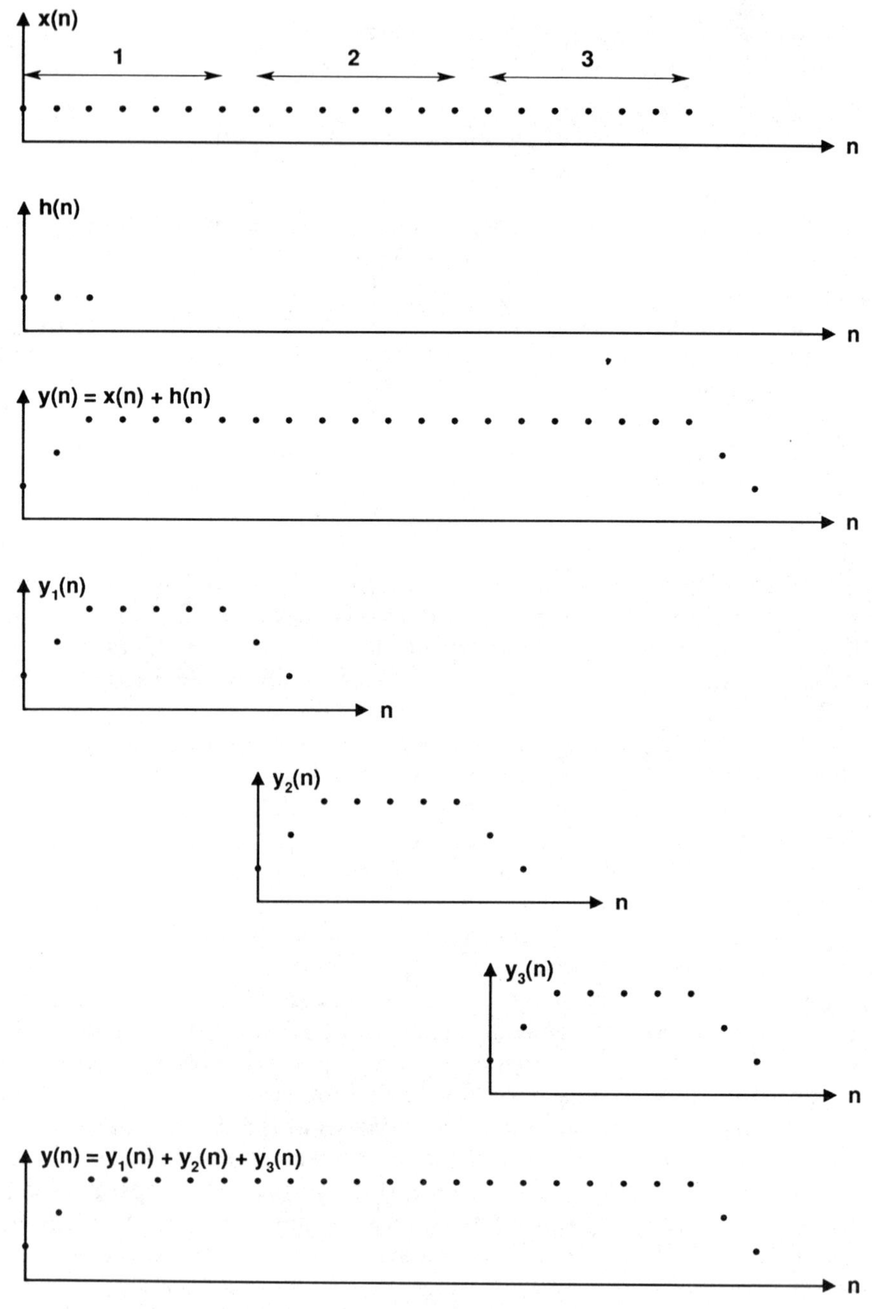

Figure 4.13 FFT filtering of long-duration signals.

EXERCISES FOR CHAPTER 4

4.1 Show that

$$x(n\,\Delta t) = \int X_{\mathrm{D}}(f)e^{i2\pi nf\,\Delta t}$$

where $X_{\mathrm{D}}(f)$ is defined in (4.3).

4.2 A low-pass filter of bandwidth B has a frequency response rect$(f/2B)$, where rect$(f/2B) = 1$ for $|f| < B$ and is zero otherwise. Show that (4.7) is the response of a low-pass filter to an input $\Delta t\,\Sigma_{n=-\infty}^{\infty} x(n\,\Delta t)\delta(t - n\,\Delta t)$.

4.3 Show that if p derivatives of $G(f)$ exist—that is, the $(p+1)$st derivative yields delta functions—then $g(t) \sim K/t^{p+1}$ for large t. Do this by writing $g(t) = \int G(f)e^{i2\pi ft}\,df$ and integrating by parts $p+1$ times.

4.4 A convenient interpolation function employs a cosine roll off, with frequency characteristics

$$G(f) = \begin{cases} 1, & |f| < W \\ \dfrac{1}{2}\left(1 + \cos\pi\,\dfrac{(f-W)}{f_{\mathrm{c}}}\right), & W \le f \le f_{\mathrm{c}} + W \\ \dfrac{1}{2}\left(1 + \cos\pi\,\dfrac{(f+W)}{f_{\mathrm{c}}}\right), & -W \ge f \ge -(f_{\mathrm{c}} + W) \\ 0, & \text{otherwise} \end{cases}$$

Show that the corresponding $g(t)$ is

$$g(t) = -\,\frac{\sin\pi t(2W + f_{\mathrm{c}})\cos\pi f_{\mathrm{c}}t}{\pi t(4f_{\mathrm{c}}^2 t^2 - 1)}$$

4.5 Verify equations (4.11) and (4.12).

4.6 By using the same criterion used in connection with (4.17) and (4.18) show that a video pulse of bandwidth B, having a rectangular spectrum, has a nominal (i.e. ~3 dB) time duration ~$1/2B$, whereas a carrier pulse of bandwidth B with a rectangular spectrum has a nominal duration $1/B$.

4.7 Prove the Riemann–Lebesgue lemma by an integration by parts in (4.27).

4.8 Prove $\displaystyle\int_{-\infty}^{\infty} \frac{\sin x}{x}\,dx = \pi$.

4.9 Show that the Hilbert transform of $\sin 2\pi f_0 t = -\cos 2\pi f_0 t$. Also, if $a(t)$ is band-limited to $|f| < B/2$ with $B \leq 2f_0$, then the Hilbert transform of $a(t)\cos 2\pi f_0 t$ is $a(t)\sin 2\pi f_0 t$ that of $a(t)\sin 2\pi f_0 t$ is $-a(t)\cos 2\pi f_0 t$.

4.10 By using (4.35) show that if $f(t)$ and $g(t)$ are band-limited to $|f| < B/2$, with $B \leq 2f_0$ that (4.36) is equal to $(\sin 2\pi f_0 t/2) f(t) * g(t)$.

4.11 If $y(t)$ is a realization of a stationary random process, band-limited to $|f| \leq B$, and correlation function $r_y(\tau) = E(y(t)y(t+\tau))$. Show that

$$r_y(\tau) = \sum_{n=-\infty}^{\infty} r_y(n/2B) \frac{\sin 2\pi B(\tau - n/2B)}{\pi(\tau - n/2B)}$$

4.12 If $f(t)$ is band-limited to $|f| \leq B$, show, using Hilbert transforms, that

$$\int f(\tau) \frac{\sin \pi 2B(t-\tau)}{\pi(t-\tau)} \, d\tau = f(t)$$

4.13 Write the Fourier transform $F(f)$ of $f(t) = P_T(t)\sin 2\pi f_1 t + P_T(t)\sin 2\pi f_2 t$, where $P_T(t) = 1$ for $0 \leq t \leq T$ and is zero otherwise. Observe that the two frequencies can be resolved—that is $|F(f)|$ will exhibit two distinct peaks—if $f_2 \geq f_1 + 1/T$. Thus the frequency resolution Δf is $1/T$.

5

DETECTION OF SIGNALS IN NOISE

Originally, the purposes of radar and sonar were (1): to determine at any instant whether or not an object of interest was present and (2): to determine its location and velocity. With the development of signal processing theory and techniques, and computers, however, active sensing applications have been greatly extended and have included weather prediction, mapping of terrain, including the ocean bottom with sonar, oil exploration, discovering ancient buried river beds, radar astronomy, and new applications are continually being found. With the development of these applications the interpretation of sensor data has of course become much more subtle and complex. Nevertheless, no matter what the application one is still concerned with the basic decision as to whether or not the data presents sufficient evidence for deciding whether or not the particular object or feature of interest—which we shall henceforth call the target—is present, and the effective signal-to-noise ratio remains an essential element for consideration in making this decision. Hence, the concepts that have been developed for dealing with the traditional radar-detection problem remain important for all active sensors for any application. In this chapter we deal with the problem of target detection. The second function, that of estimating target parameters such as position and velocity is dealt with in Chapter 7.

5.1 STATISTICAL DECISION THEORY: THE LIKELIHOOD RATIO TEST

Very early in the development of radar it was recognized that the ultimate limitation on the performance of a radar resides in the presence of the random processes affecting it. This holds true for all active sensing systems since the observable is always a random variable, and the detection problem is therefore fundamentally statistical. One must deal with observables $y(t)$ of the form

$$y(t) = y[s(t), n(t)] \tag{5.1}$$

where $s(t)$ is a signal of interest and $n(t)$ is a realization of some random process.

Statistical decision theory provides a framework within which the detection problem for sensing systems can be formulated. We speak of two hypothesis: H_1, that a target is present, and H_0, that a target is not present. H_0 is referred to as the null hypothesis, and H_1 as the alternative. This formulation is equally applicable to a binary digital communication system in which each transmitted symbol is either a 1 or a 0, which corresponds to H_1 and H_0 respectively.

In general, the problem can be formulated in terms of a null hypothesis and a compound alternative, which would apply for example to an M-ary communication system. We do not deal with this here.

We wish to maximize the probability of a correct decision $P[C]$ which can be expressed as

$$P[C] = \int P[C|y]P(y)\,dy \tag{5.2}$$

where $P[C|y]$ is the conditional probability of a correct decision, given the particular observable $y(t)$ which is received. From (5.2) it is clear, since $P(y) \geq 0$, that, for any given received $y(t)$, $P[C]$ is maximized by making that decision which maximizes $P[C|y]$. Therefore, the ideal decision scheme is that a target shall be declared present if and only if (iff)

$$P(H_1|y) > P(H_0|y) \tag{5.3}$$

That is, the conditional probabilities $P(H_1|y)$ and $P(H_0|y)$ will vary depending on the value of the observable $y(t)$. If, say, H_1 is chosen then $P(C|y) = P(H_1|y)$. Hence for any given value of $y(t)$ the unconditional probability of correct decision $P(C)$ is maximized by choosing H_1 or H_0 depending on whether $P(H_1|y)$ or $P(H_0|y)$ is larger. Putting it another way, $P(H_1|y) > P(H_0|y)$ would certainly not be a good reason for deciding that a target is not present.

In order to evaluate (5.3), apply Baye's Theorem, (2.20), and the rule becomes, choose H_1 if

$$\frac{P(H_1)P(y|H_1)}{P(y)} > \frac{P(H_0)P(y|H_0)}{P(y)} \tag{5.4}$$

or

$$\frac{P(y|H_1)}{p(y|H_0)} > \frac{P(H_0)}{P(H_1)} = T \tag{5.5}$$

The functions $P(y|H_1)$ and $P(y|H_0)$ are the a priori conditional probabilities for the events H_0 and H_1. In contrast with the a posteriori conditional probabilities $P(H_0|y)$ and $P(H_1|y)$, they can be calculated from knowledge of the noise statistics before any observations are made, which is the essential feature that makes this approach useful for defining detection criteria. However, the ideal Bayes decision scheme, which maximizes the probability of a correct decision, requires knowledge not only of the noise statistics, but also of the a priori probabilities of H_0 and H_1. Let us hold off discussion of this important consideration for the moment.

$P(y|H_0)$ and $P(y|H_1)$ are also referred to as likelihood functions, the function $P(y|H_1)/P(y|H_0)$ is referred to as a likelihood ratio, and (5.5) is a particular form of a general decision scheme known as a likelihood-ratio test (LRT), in which the decision regarding the choice of H_0 or H_1 is made dependent on whether the likelihood ratio is greater or less than a certain predetermined threshold T. The prescription for a LRT is therefore as follows

1. In a manner to be discussed shortly calculate the threshold T.
2. From knowledge of the noise statistics calculate the likelihood functions $P(y|H_1)$ and $P(y|H_0)$. Referring to Figure 5.1 it is seen that

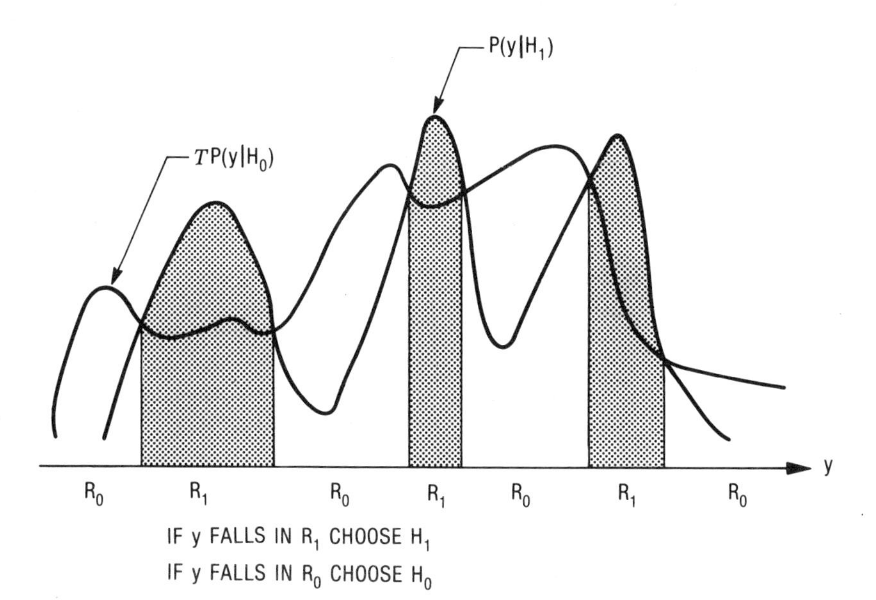

Figure 5.1 Decision regions.

these two steps define the decision regions R_0 and R_1 in which $P(y|H_1)$ is, respectively, less than and greater than $TP(y|H_0)$.

3. Make an observation of y. If y falls in R_0 decide on H_0 and if y falls in R_1 decide on H_1. Since in practice decisions are often based on several observations the case for y exactly on a boundary need not realistically be considered.

5.2 DECISION CRITERIA: BAYES, MAXIMUM LIKELIHOOD, NEYMAN–PEARSON

The decision criterion determines the value of the threshold T as follows:

1. *Bayes Criterion*. As noted, the optimum choice, from the point of view of maximizing the probability of a correct decision, is to let $T = P(H_0)/P(H_1)$. But of course $P(H_0)$ and $P(H_1)$ will in general not be known a priori which is an obvious problem for the Bayes scheme.

2. *Maximum Likelihood Criterion*. This scheme considers simply which of two likelihood functions $P(y|H_0)$ and $P(y|H_1)$ is larger. That is, the threshold T is unity, and H_1 is selected if $P(y|H_1)$ is greater than $P(y|H_0)$, and vice versa. This is equivalent to the Bayes scheme for $P(H_0) = P(H_1) = \frac{1}{2}$, and is applicable to a binary communication system in which it can be assumed that any given symbol will equally likely be a 1 or a 0.

3. *Neyman–Pearson Criterion*. Both the Bayes and Maximum Likelihood schemes suffer from the shortcoming that even with the introduction of cost functions, which are not considered here, it is difficult to make a quantitative evaluation of the effectiveness of the decision scheme since the cost functions themselves are most often either unknown or arbitrary. The Neyman–Pearson approach on the other hand yields a decision scheme subject to a tangible quantitative evaluation.

Referring to Figure 5.1, for any choice of T, which determines the boundaries of the decision regions R_0 and R_1, there will always be a finite probability that H_0 will be true but y will fall into R_1 or, conversely, H_1 will be true and y will fall into H_0. In either event the decision scheme will yield the incorrect answer, and the associated errors are defined as:

Type 1 Error—The probability α that H_1 is selected but H_0 is true, given by

$$\alpha = \int_{R_1} P(y|H_0)\, dy \tag{5.6}$$

Type 2 Error—The probability β that H_0 is selected but H_1 is true, given by

$$\beta = \int_{R_0} P(y|H_1)\, dy \tag{5.7}$$

In binary communication systems since it is equally likely that any given symbol will be a 1 or a 0 (or +1 or −1) the Bit Error Rate (BER) is

$$P(\text{a 1 is sent})\ P(\text{Error}|\text{1 is sent}) + P(\text{a 0 is sent})\ P(\text{Error}|\text{0 is sent}) = \frac{\alpha + \beta}{2}$$

In sensing systems the Type 1 error is commonly referred to as the false alarm probability P_{fa}, because it is the probability of declaring a target to be present when in fact it is not. Rather than dealing directly with the Type 2 error however, it is more customary to deal with the probability of correctly declaring a target to be present—the detection probability P_{d}—which is given by

$$P_{\text{d}} = \int_{R_1} P(y|H_1)\, dy = 1 - \int_{R_0} P(y|H_1)\, dy = 1 - \beta \tag{5.8}$$

Equation (5.8) holds because in either case we must of course have

$$\begin{aligned} \int_{R_0} P(y|H_0)\, dy + \int_{R_1} P(y|H_0)\, dy = 1 \\ \int_{R_0} P(y|H_1)\, dy + \int_{R_1} P(y|H_1)\, dy = 1 \end{aligned} \tag{5.9}$$

In the Neyman–Pearson approach first a value of P_{fa} is selected. How this might be done is discussed in Chapter 11. Obviously it depends on what action will be taken if a target is declared to be present and how often one can afford to take this action needlessly. Given this value of $P_{\text{fa}} = \alpha$, the boundaries of R_0 and R_1—or equivalently the threshold T—are selected in order to maximize P_{d}. The problem therefore is that of maximizing P_{d} subject to the constraint $P_{\text{fa}} = \alpha$, which can be formulated as maximizing the quantity $\mathscr{D}$, given by

$$\mathscr{D} = P_{\text{d}} + \eta(\alpha - P_{\text{fa}}) \tag{5.10}$$

where η is a Lagrange multiplier. By definition, (5.10) is

$$\begin{aligned} \mathscr{D} &= \int_{R_1} P(y|H_1)\, dy + \eta\left[\alpha - \int_{R_1} P(y|H_0)\, dy\right] \\ &= \eta\alpha + \int_{R_1} [(P(y|H_1) - \eta P(y|H_0)\, dy] \end{aligned} \tag{5.11}$$

and $\mathscr{D}$ is maximized by choosing R_1 such that the integrand in (5.11) is positive. The decision rule is therefore choose H_1 if

$$\frac{P(y|H_1)}{P(y|H_0)} > \eta \tag{5.12}$$

and consequently choose H_0 if

$$\frac{P(y|H_1)}{P(y|H_0)} < \eta \tag{5.13}$$

Hence the Lagrange multiplier η turns out to be the decision threshold, which now replaces T. To evaluate η let $\Lambda(y) = P(y|H_1)/P(y|H_0)$. Then η is determined by the constraint α from

$$\alpha = \int_{\eta}^{\infty} P[\Lambda(y)|H_0]\, dy \tag{5.14}$$

Equation (5.14) is the formal expression for η, which shall be elaborated on shortly.

It has therefore been proved that P_d is maximized, subject to the constraint $P_{fa} = \alpha$, when the decision region, or equivalently the threshold η, is chosen according to (5.14); this result is known as the Neyman–Pearson lemma. The Neyman–Pearson criterion therefore has an advantage over the Bayes criterion in that although the latter yields the maximum probability of a correct decision, there is no control over the false alarm probability, which may turn out to be too large and therefore too high a price to pay for a maximum probability of being correct.

In summary, regardless of the particular decision criterion, in order to implement an LRT we:

1. Calculate $\Lambda(y)$ under H_0 and H_1 from knowledge of the noise statistics and the form of the observable $y(t)$.
2. For some physical observation y_0 evaluate $\Lambda(y_0)$.
3. Compare $\Lambda(y_0)$ with a prescribed threshold η. It is only in the value of η that the particular decision criterion manifests itself; the form of the decision procedure remains the same in all cases. For the Bayes rule, $\eta = P(H_0)/P(H_1)$ and the LRT maximizes the probability of a correct decision. For maximum likelihood $\eta = 1$ and the LRT yields the maximum probability of a correct decision provided $P(H_0) = P(H_1) = \frac{1}{2}$, which of course may or may not be true. For the Neyman–Pearson approach η is chosen to correspond to the desired value of P_{fa}. In this case the LRT maximizes the probability of a correct decision subject to this constraint on P_{fa}.

5.3 IMPLEMENTATION OF DECISION CRITERIA

The foregoing results are completely general. The continuance of this development shall be restricted to the two most important forms of statistical interference, namely, additive, mean-zero white, Gaussian noise and shot noise, which has a Poisson distribution. As is discussed in Chapter 1, we do not deal with randomness in the observables contributed by clutter, reverberation, random propagation phenomena, or fluctuations in target scattering characteristics.

The ideal Gaussian channel is of greatest relevance for radar, many types of communications channels, and also sonar, in which the white-Gaussian assumption for ambient ocean noise is also reasonable. Nonwhite or colored noise is generally dealt with by employing prewhitening stages in the processing path, and while this complicates the processing operations it does not alter the fundamental approach. Optical systems such as laser radar, as has been discussed in Chapter 3, are subject to both thermal and shot noise.

5.3.1 Gaussian Noise

Since the observables are time functions, it is necessary to deal with random processes and multiple observations. To begin with however let us consider single observations of random variables. Let the observable y under the two hypothesis be given by

$$\begin{aligned} &\text{under } H_0 \qquad y = n \\ &\text{under } H_1 \qquad y = s + n \end{aligned} \tag{5.15}$$

where s is a deterministic quantity and n is a mean-zero Gaussian random variable with probability density function (pdf).

$$P_n(x) = \frac{e^{-x^2/2\sigma^2}}{\sqrt{2\pi}\sigma} \tag{5.16}$$

The conditional pdfs for the random variables y and n are related by $P(y|H_0) = P_n(y|H_0)$, $P(y|H_1) = P_n(y - s|H_1)$, and the likelihood ratio test is therefore, select H_1 if

$$\frac{P(y|H_1)}{P(y|H_0)} = \frac{P_n(y - s|H_1)}{P_n(y|H_0)} = \frac{e^{-(y-s)^2/2\sigma^2}}{e^{-y^2/2\sigma^2}} > \eta \tag{5.17}$$

After taking the logarithm of both sides of (5.17) the criterion for selection of H_1 is

$$y > \frac{\sigma^2}{s} \log \eta + \frac{s}{2} \tag{5.18}$$

In (5.18) it is seen that the LRT, in which $\Lambda(y)$ of (5.12) is compared with a threshold η, becomes transformed into a comparison of the observable y with the threshold in (5.18), which is a function of η. As an example, suppose $P(H_0)$ and $P(H_1)$ are known, with $P(H_0)/P(H_1) = 2$. Then the Bayes and maximum likelihood decision rules are, choose H_1 if

Bayes

$$y > \frac{s}{2} + \frac{\sigma^2 \log 2}{s} \tag{5.19}$$

Maximum Likelihood ($\eta = 1$)

$$y > \frac{s}{2}$$

Since the a priori probability of H_0 is twice that of H_1, then the Bayes rule requires a larger value of y for selection of H_0 than the maximum likelihood rule, in which this information is not used. The Bayes scheme therefore of course yields a better decision rule in this case.

For the Neyman–Pearson criterion, suppose a value of $P_{\text{fa}} = 0.01$ can be tolerated. The threshold is then that value of η for which

$$P_{\text{fa}} = 0.01 = \int_{\eta}^{\infty} e^{-x^2/2\sigma^2} \frac{dx}{\sqrt{2\pi}\sigma} = \frac{1}{2} - \int_{0}^{\eta} e^{-x^2/2\sigma^2} \frac{dx}{\sqrt{2\pi}\sigma}$$
$$= \frac{1}{2}\left[1 - \operatorname{erf}\left(\frac{\eta}{\sqrt{2}\sigma}\right)\right] \tag{5.20}$$

where the error function erf(x) is

$$\operatorname{erf}(x) = \frac{2}{\sqrt{\pi}} \int_{0}^{x} e^{-t^2}\, dt \tag{5.21}$$

Thus, equation (5.14), which establishes the threshold condition on $\Lambda(y)$, becomes transformed in terms of the observable y into

$$\text{choose } H_1 \text{ if } y > \eta \tag{5.22}$$

From the table of erf(x) in the Appendix, for $P_{\text{fa}} = 0.01$ it is required that $\eta/\sqrt{2}\sigma = 1.646$ or, referring to (5.18)

$$\eta = \sqrt{2}(1.65)\sigma = 2.33\sigma$$

A typical illustration of the thresholds for the three decision criteria for this example is given in Figure 5.2.

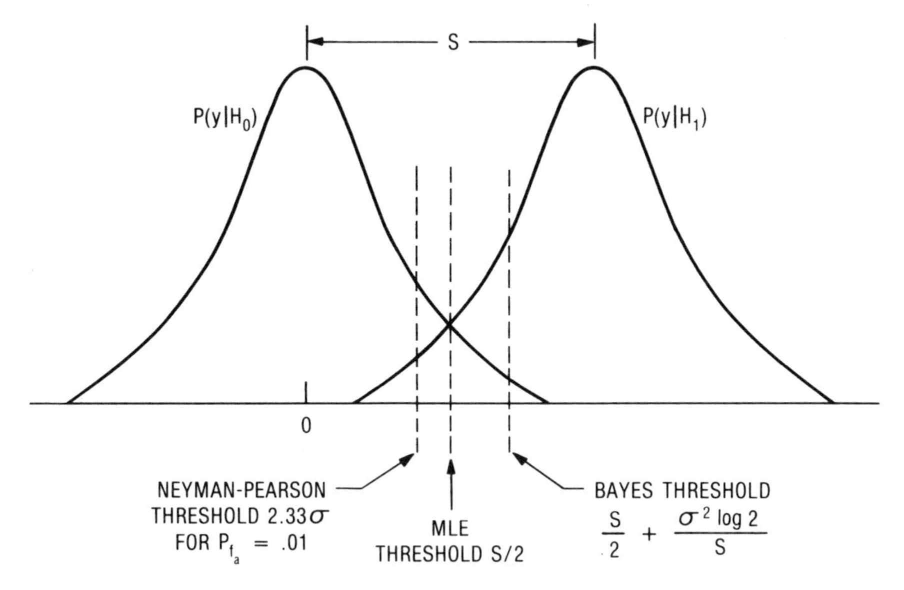

Figure 5.2 Decision thresholds.

Returning to the Neyman–Pearson case, the detection probability is

$$\begin{aligned} P_{\mathrm{d}} &= \int_{\eta}^{\infty} e^{-(y-s)^2/2\sigma^2} \frac{dy}{\sigma\sqrt{2\pi}} \\ &= \frac{1}{2}\left[1 + \mathrm{erf}\left(\frac{s-\eta}{\sqrt{2}\sigma}\right)\right] \\ &= \frac{1}{2}\left[1 + \mathrm{erf}\left(\frac{s}{\sqrt{2}\sigma} - 1.646\right)\right] \end{aligned} \tag{5.23}$$

and P_{d} depends only on the signal-to-noise ratio (SNR)[†], through $s/\sigma\sqrt{2}$, and of course also on the assigned value of P_{fa}, which determines the numerical parameter, 1.646 in this example. To summarize (see Ex. 5.1)

Bayes

$$P_{\mathrm{fa}} = \frac{1}{2}\left[1 - \mathrm{erf}\left(\frac{s}{2\sigma\sqrt{2}} + \frac{\sigma \log \eta}{s\sqrt{2}}\right)\right]$$

$$P_{\mathrm{d}} = \frac{1}{2}\left[1 + \mathrm{erf}\left(\frac{s}{2\sigma\sqrt{2}} - \frac{\sigma \log \eta}{s\sqrt{2}}\right)\right]$$

[†] SNR will be defined in Sec. (5.5).

Maximum Likelihood

$$P_{\mathrm{fa}} = \left[1 - \mathrm{erf}\left(\frac{s}{2\sigma\sqrt{2}}\right)\right]$$

$$P_{\mathrm{d}} = \left[1 + \mathrm{erf}\left(\frac{s}{2\sigma\sqrt{2}}\right)\right]$$

Neyman–Pearson

$$P_{\mathrm{fa}} = \int_{\eta}^{\infty} e^{-x^2/2\sigma^2} \frac{dx}{\sqrt{2\pi}\sigma} = \frac{1}{2}\left(1 - \mathrm{erf}(\gamma)\right)$$

where $\gamma = \eta/\sigma\sqrt{2}$ and

$$P_{\mathrm{d}} = \frac{1}{2}\left[1 + \mathrm{erf}\left(\frac{s}{\sqrt{2}\sigma} - \gamma\right)\right]$$

In all cases, since $\mathrm{erf}(x) \rightarrow \frac{1}{2}$ for large x, P_{d} can be made arbitrarily close to unity for large enough signal-to-noise ratio, which is proportional to s^2/σ^2, and which arises because of the mathematical form of the Gaussian distribution. The three criteria however embody certain fundamental differences. For Maximum Likelihood the threshold is such that $P_{\mathrm{d}} = 1 - P_{\mathrm{fa}}$ which does not hold in the other two cases. For the Bayes criterion the parameter $\eta = P(y/H_1)/P(y/H_0)$ is such that the probability of a correct decision is maximized. But even if $P(y/H_1)$ and $P(y/H_0)$ were known, this might not be the most desirable choice, because it could result in an unacceptably large false-alarm probability, over which the Bayes scheme has no control. On the other hand, with the Neyman–Pearson approach P_{fa} can be specified to any desired value, in which case the maximum value of P_{d} is achieved subject to this constraint.

5.3.2 Shot Noise—Poisson Distribution

A very simplified diagram of an optical receiver is shown in Figure 5.3. Optical receivers essentially count photons. The photo electric current at the output of the detector is proportional to the power in the incident light, the

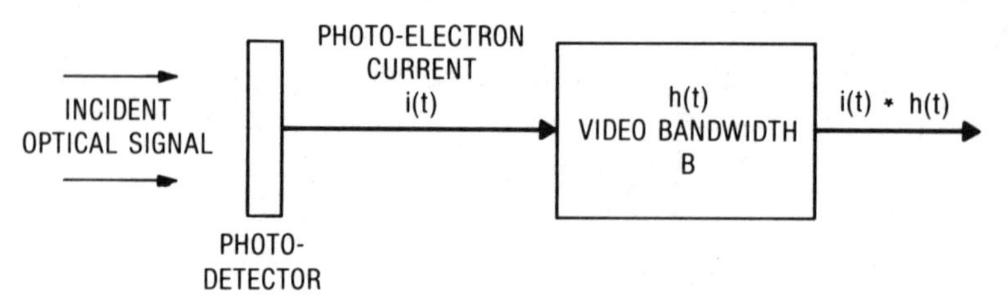

Figure 5.3 Typical optical receiver.

proportionality constant being the optical efficiency of the detector. A band-limiting of the inherent impulsive property of the individual photoelectrons is represented by the baseband filter $h(t)$ in Figure 5.3. If there is no actual filter in the path, then $h(t)$ represents the band-limiting imposed by the finite bandwidth of the optical detector itself.

As is discussed in Section 4.3, the duration of each video current pulse is nominally $\tau = 1/2B$, where B is the bandwidth of $h(t)$ (τ as used here has nothing to do with the arrival-time usage of τ in Section 3.4). The output of the filter is of the form of a convolution of $h(t)$ with the photo-electron current $i(t)$, and at any time t therefore contains contributions from all the current pulses which arrived during the interval $t - \tau$. The output is therefore a measure of the number of pulses, and hence also the number of photons, which arrived during this interval. This observable is denoted as the count function $k(t)$ and is a random process in which, for all i and j, the random variables $k(t_i)$ and $k(t_j)$ are independent for $|t_j - t_i| \geq \tau$, which follows from the foregoing discussion.

In an optical system, even if the lenses are capped there will generally always be some form of optical radiation incident on the detector due to the finite temperature of various system components such as baffles, the telescope barrel, etc. Also, when operating, there may be optical radiation present whose strength will depend on the temperature of the background against which the targets of interest are viewed. The current which is produced by all such ambient radiation when no signal is present is called the dark current i_d. Hence, if i is the photoelectron current, the average number of photoelectron pulses per second under the two hypotheses will be

$$
\begin{aligned}
H_0:&\quad i/q = i_d/q \\
H_1:&\quad i/q = (i_d + i_s)/q
\end{aligned}
\tag{5.24}
$$

where i_s is the current produced by the optical signal of interest and q is the electronic charge.

Now, referring to (2.11), the probability of k events taking place in T seconds is

$$
P(k) = e^{-\lambda T} \frac{(\lambda T)^k}{k!} \tag{5.25}
$$

where λ is the average number of events per second; as is discussed in Chapter 2, if λT is large enough the distribution can be approximated by a Gaussian. The events in this case are the occurrence of current pulses, for which the rates of occurrence are i_d/q and $(i_d + i_s)q$. But, as noted above, each observation $k(t)$ is a measure of the number of current pulses which occurred during the previous $\tau = 1/2B$ seconds, and the relevant quantity is therefore the number of counts during this interval. Using (5.25) the LRT is therefore: declare H_1 if

$$\frac{P(k|H_1)}{P(k|H_0)} = \frac{[(i_d + i_s)(\tau/q)]^k \exp[-(i_d + i_s)(\tau/q)]}{(i_d \tau/q)^k \exp(-i_d \tau/q)}$$

$$= \left(1 + \frac{i_s}{i_d}\right)^k \exp\left(\frac{-i_s \tau}{q}\right) > \eta \tag{5.26}$$

and, equivalently, in terms of the random variable k, declare H_1 if

$$k > \frac{\log \eta}{\log[1 + (i_s/i_d)]} + \frac{i_s \tau}{q} \tag{5.27}$$

With the parameter η equal to unity, the maximum likelihood rule is to declare H_1 if k exceeds the expected average number of counts during the interval τ due to signal alone.

For the Neyman–Pearson criterion use

$$P(k|H_0) = \frac{\exp[-i_d(\tau/q)]}{k!} \left(i_d \frac{\tau}{q}\right)^k \tag{5.28}$$

and for the specified value of P_{fa} determine that value k_0 which satisfies

$$P_{fa} = \sum_{n=k_0}^{\infty} \frac{\exp[-i_d(\tau/q)]}{n!} \left(\frac{i_d \tau}{q}\right)^n \tag{5.29}$$

Determination of detection thresholds is somewhat more cumbersome than for the Gaussian case. The basic approach however remains unchanged. The rule is: declare H_1 if $k > k_0$. The detection probability in this case is therefore

$$P_d = \sum_{k=k_0}^{\infty} \frac{\exp[-(i_s + i_d)\tau/q]}{k!} \left[(i_s + i_d) \frac{\tau}{q}\right)^k \tag{5.30}$$

Let k_s, k_d and ρ denote respectively the number of counts during the interval τ due to signal, dark current and signal plus dark current. These are given by

$$k_s = \frac{i_s}{2qB} \qquad k_d = \frac{i_d}{2qB}$$
$$\rho = k_s + k_d = \frac{i_d + i_s}{2qB} \tag{5.31}$$

And the foregoing results can be summarized as: declare H_1 if the observed count satisfies:

Bayes

$$k > \frac{\log(P(H_0)/P(H_1))}{\log(1 + k_s/k_d)} + k_s \tag{5.32}$$

Maximum Likelihood

$$k > k_s \tag{5.33}$$

Neyman–Pearson

Specify P_{fa} and determine the threshold k_0 for which

$$P_{fa} = \sum_{n=k_0}^{\infty} \frac{e^{-k_d} k_d^n}{n!} = 1 - \sum_{n=0}^{k_0-1} \frac{e^{-k_d} k_d^n}{n!} \tag{5.34}$$

In which case the detection probability is

$$P_d = 1 - \sum_{n=0}^{k_0-1} \frac{e^{-\rho} \rho^n}{n!} \tag{5.35}$$

and of course declare H_0 if $k < k_0$ and H_1 if $k > k_0$.

In the foregoing results the dark current i_d plays the role that σ^2 plays in the Gaussian case in Section 5.3.1. That is, signal detection contends with the fluctuations produced by the random arrival times of the photons which produce i_d as well as those in the signal. Hence, as for σ^2 in the Gaussian case, here i_d must be known in order to specify thresholds, etc.; also, if $i_d = 0$ then detection of a single photon would denote the presence of the signal of interest and, with the Neyman–Pearson criterion, $P_{fa} = 0$, $P_d = 1$. In the Gaussian case this would correspond to $\sigma^2 = 0$.

5.4 CORRELATION DETECTION: THE MATCHED FILTER—I

The previous examples dealt with single observations of a random variable. In practice however random processes must be considered, because the observables are generally time functions and one must therefore deal with multiple observations taking place over an interval equal to the signal duration. The foregoing analysis shall now be extended to this situation.

5.4.1 The Gaussian Channel

Let the signal of interest $s(t)$ be of duration T, and let the interference consist of mean-zero white Gaussian noise $n(t)$ with two-sided power spectral density $N_0/2$. It is to be emphasized that $s(t)$ is perfectly general and

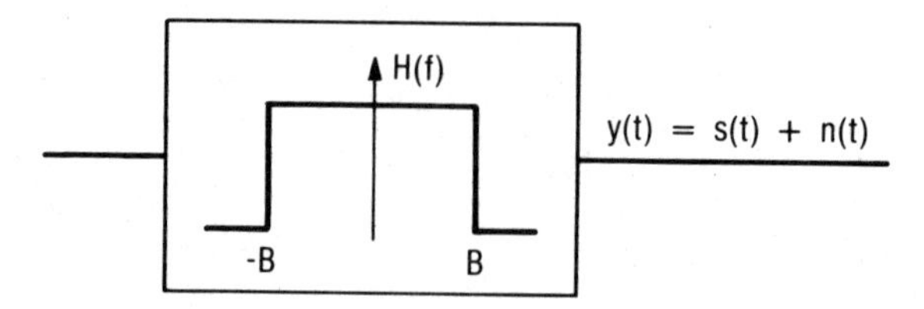

Figure 5.4 Observations at output of low-pass filter.

the following results therefore apply to all cases of interest, including in particular, carrier as well as video waveforms. Let the observation take place at the output of a hypothetical ideal low-pass filter of bandwidth B (Figure 5.4) which is sufficiently wide to pass all of the signal energy; subsequently, B shall be allowed to become infinite. Then the noise becomes limited in frequency to the range $-B \leq f \leq B$, and by the Wiener–Khinchine theorem, the correlation function $r(\tau)$ of $n(t)$ is

$$r(\tau) = \frac{N_0}{2} \int_{-B}^{B} e^{i2\pi f\tau}\, df = \frac{N_0 B}{2} \frac{\sin 2\pi B\tau}{\pi B\tau} \tag{5.36}$$

Since $r(\tau) = 0$ for $\tau = n/2B$, $n = 0, 1, 2, 3, \ldots$, samples of $n(t)$ separated in time by $1/2B$ seconds will be independent. Therefore, if an observation takes place over T seconds, with sampling at the foregoing rate, if the signal $s(t)$ is present there will be $2BT$ independent random variables $y(t_i) = s(t_i) + n(t_i)$, $i = 1, 2, \ldots, 2BT$. Therefore, since the joint density function of N independent random variables is the product of the N density functions (2.18), the joint probability density of the observation in terms of the random variables $n(t_i)$ is

$$\begin{aligned} P(n(t_1), n(t_2), \ldots) &= \prod_{i=1}^{2BT} \frac{\exp[-(y(t_i) - s(t_i))^2/2\sigma^2]}{(2\pi)^{BT}\sigma^{2BT}} \\ &= \frac{\exp\left[-\frac{1}{2\sigma^2} \sum_{i=1}^{2BT} (y(t_i) - s(t_i))^2\right]}{(2\pi)^{BT}\sigma^{2BT}} \end{aligned} \tag{5.37}$$

where $\sigma^2 = N_0 B$ is the noise power. If the signal is not present the observation will of course consist of noise alone and the LRT in this case—the generalization of (5.17)—becomes, choose H_1 if

$$\frac{\exp\left[-\frac{1}{2\sigma^2} \sum_{i=1}^{2BT} (y(t_i) - s(t_i))^2\right]}{\exp\left[-\frac{1}{2\sigma^2} \sum_{i=1}^{2BT} y^2(t_i)\right]} = \exp\left\{\frac{1}{2\sigma^2} \sum_{i=1}^{2BT} [2y(t_i)s(t_i) - s^2(t_i)]\right\} > \eta \tag{5.38}$$

or equivalently if

$$\frac{1}{\sigma^2}\sum_{i=1}^{2BT} y(t_i)s(t_i) > \log \eta + \frac{1}{2\sigma^2}\sum_{i=1}^{2BT} s^2(t_i) \tag{5.39}$$

The decision rule can be expressed in terms of continuous functions by making use of eq. (4.7), that an arbitrary function $u(t)$ whose energy is contained within a frequency range B—that is, a function band-limited to a range B—can be written as

$$u(t) = \sum_{u=-\infty}^{\infty} u\left(\frac{n}{2B}\right) \frac{\sin 2\pi B(t - n/2B)}{2\pi B(t - n/2B)}$$

where in this case it is assumed that sampling takes place at the Nyquist rate $1/\Delta t = 2B$. It then follows that for any two band-limited functions $u(t)$ and $v(t)$ we can write

$$\int_{-\infty}^{\infty} u(t)v(t)\,dt = \sum_n \sum_m u\left(\frac{n}{2B}\right) v\left(\frac{m}{2B}\right) \int_{-\infty}^{\infty} \frac{\sin 2\pi B(t - n/2B)}{2\pi B(t - n/2B)}$$

$$\frac{\sin 2\pi B(t - m/2B)}{2\pi B(t - m/2B)}\,dt$$

$$= \frac{1}{2B}\sum_{n=-\infty}^{\infty} u\left(\frac{n}{2B}\right) v\left(\frac{n}{2B}\right) \tag{5.40}$$

because the set of functions $\sin \pi(x-n)/\pi(x-n)$ are orthonormal on the index n over the interval $-\infty \le x \le \infty$ see Exercise 5.3). As an auxiliary result,

$$\int_{-\infty}^{\infty} u^2(t)\,dt = E = \frac{1}{2B}\sum_{n=-\infty}^{\infty} u^2\left(\frac{n}{2B}\right) \tag{5.41}$$

where E is the signal energy.

As is discussed in Chapter 4 in connection with (4.7), the foregoing results, which are exact for band-limited functions, are only approximate for functions of finite duration. For any value of T however, the approximation improves as the number of terms in the summation $T/\Delta t = 2BT$ increases. Therefore, let B become arbitrarily large so that $\Delta t \to 0$, which amounts to continuous sampling, and to an arbitrarily good approximation the criterion for choice of H_1, (5.38), is, by (5.40),

$$\exp\left[\frac{1}{N_0}\int_0^T (2y(t)s(t) - s^2(t)\,dt\right] > \eta$$

and (5.39) becomes

$$\frac{2}{N_0}\int_0^T y(t)s(t)\,dt > \log\eta + \frac{E}{N_0} \tag{5.42}$$

where $\sigma^2 = N_0B$. Stated another way, a time function of finite duration has infinite bandwidth, and since $B \to \infty$ the solution becomes exact in the limit. The left-hand side becomes, under the two hypotheses

$$\begin{aligned} H_0&: \quad \frac{2}{N_0}\int_0^T y(t)s(t)\,dt = \frac{2}{N_0}\int_0^T n(t)s(t)\,dt \\ H_1&: \quad \frac{2}{N_0}\int_0^T y(t)s(t)\,dt = \frac{2E}{N_0} + \frac{2}{N_0}\int_0^T n(t)s(t)\,dt \end{aligned} \tag{5.43}$$

which shall be used shortly. As before, the foregoing discussion in Section 5.3.1 concerning the choice of the threshold and the various decision criteria holds here as well. The observable however is a weighted integral of the received quantity rather than the received quantity itself.

This decision rule can be implemented in a receiver such as is illustrated in Figure 5.5. It is assumed that an *exact* replica of the received signal is available which is correlated with the observable $y(t)$. In this case it would either be necessary to know exactly when $y(t)$ arrives, which of course is not known a priori, or it would be necessary to continually repeat the process of multiplication and integration ad infinitum. In fact, in discrete-time systems, where computers are available for signal processing, correlation detection is often implemented in just this way, with successive trials being separated by a sampling interval; in this case the sampling rate must be significantly greater than the Nyquist rate. An alternative approach however, in continuous time, is as follows. Recall (see Section 4.5) that for a filter, or any linear system with physically realizable (causal) impulse response $h(t)$, the output $x(t)$ for an input $y(t)$ is

$$x(t) = \int_0^t h(\tau)y(t-\tau)\,d\tau \tag{5.44}$$

Therefore let the filter have impulse response

$$h(t) = s(T-t) \tag{5.45}$$

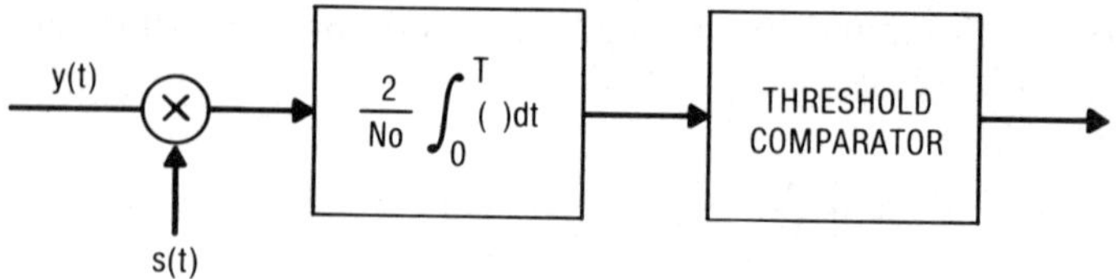

Figure 5.5 Correlation receiver.

Then for an input $y(t)$ the output $x(t)$ under the two hypotheses is

$$H_0:\quad x(t) = \int_0^t s(T-\tau)y(t-\tau)\,d\tau = \int_0^t s(T-\tau)n(t-\tau)\,d\tau$$

$$H_1:\quad x(t) = \int_0^t s(T-\tau)y(t-\tau)\,d\tau = \int_0^t s(T-\tau)s(t-\tau)\,d\tau + \int_0^t s(T-\tau)n(t-\tau)\,d\tau \tag{5.46}$$

which, at the time of maximum response, $t = T$, is identical to the output of the correlation detector equation (5.43)

$$H_0:\quad x(T) = \int_0^T s(T-\tau)n(T-\tau)\,d\tau = \int_0^T s(t)n(t)\,dt = x_n(T)$$

$$\begin{aligned} H_1:\quad x(T) &= \int_0^T s(T-\tau)s(T-\tau)\,d\tau + \int_0^T s(T-\tau)n(T-\tau)\,dt \\ &= \int_0^T s^2(t)\,dt + \int_0^T s(t)n(t)\,dt \\ &= x_s(T) + x_n(T) \end{aligned} \tag{5.47}$$

Thus the repetitive active operation illustrated in Figure 5.5 is replaced by the continuous passive operation represented by (5.44), since the filter $h(t)$ is always ready to receive a signal. No a priori information is required. If the threshold is exceeded it is assumed that correlation has occurred, at time $t = T$ in (5.46) and (5.47), and H_1 is declared. Of course there will be times when the threshold is exceeded and, in fact, no signal is present, but this is accounted for in the setting of the threshold in accordance with the specified false alarm and detection probabilities.

Equation (5.45) defines the impulse of the celebrated matched filter [18], which is the time reversed and suitably time shifted replica of the signal of interest, the time shift being required for filter causality. An example of a signal and the corresponding matched-filter impulse response is illustrated in Figure 5.6. The output of a matched filter is in units of energy.

To calculate false alarm and detection probabilities, referring to (5.42) and (5.43), it will be convenient in what follows to write the output of the matched filter under the two hypotheses as a dimensionless quantity, normalized with respect to the noise power spectral density, as:

$$\text{under } H_1:\quad x = \frac{2}{N_0}\int_0^T y(t)s(t)\,dt = \frac{2E}{N_0} + \frac{2}{N_0}\int_0^T n(t)s(t)\,dt$$

$$\text{under } H_0:\quad x = \frac{2}{N_0}\int_0^T y(t)s(t)\,dt = \frac{2}{N_0}\int_0^T n(t)s(t)\,dt \tag{5.48}$$

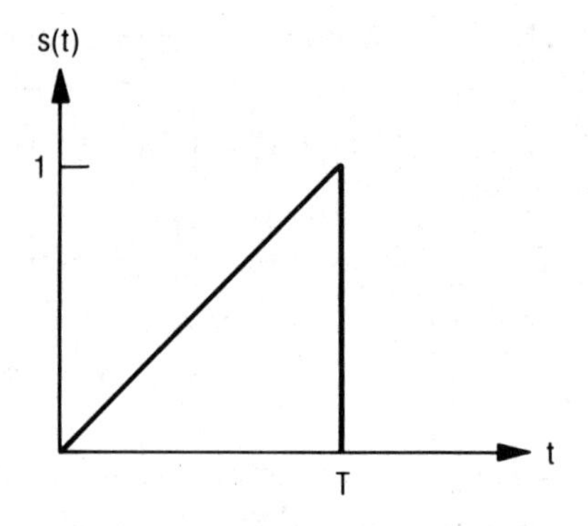

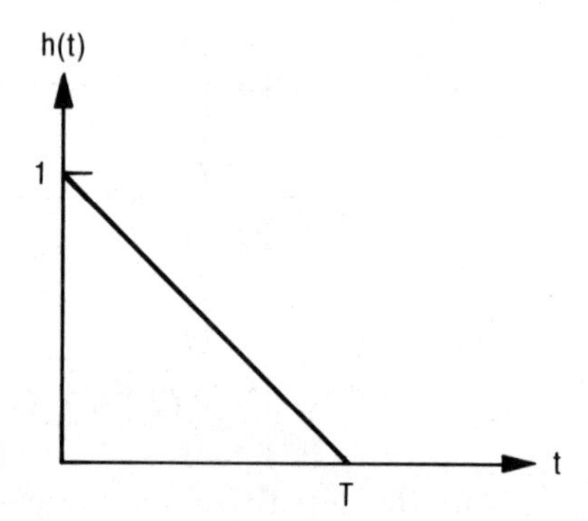

Figure 5.6 Signal, and corresponding causal matched-filter impulse response.

Now $(2/N_0)\int_0^T n(t)s(t)\,dt$ is a mean-zero Gaussian random variable with variance equal to

$$E\left[\frac{4}{N_0^2}\int_0^T n(x)s(x)\,dx \int_0^T n(y)s(y)\,dy\right]$$
$$= \frac{4}{N_0^2}\int_0^T\int_0^T E[n(x)n(y)]s(x)s(y)\,dx\,dy = \frac{2E}{N_0} \tag{5.49}$$

since for white noise the correlation function is

$$E[n(x)n(y)] = \frac{N_0}{2}\,\delta(x-y) \tag{5.50}$$

The false alarm probability is therefore

$$P_{\text{fa}} = \int_\eta^\infty e^{-x^2/2(2E/N_0)} \frac{dx}{(2\pi(2E/N_0))^{1/2}} = \frac{1}{2}\,(1-\text{erf}(\gamma)) \tag{5.51}$$

which assigns a value to the parameter $\gamma = \eta/(2\sqrt{E/N_0})$; erf is the error function defined in (5.21).

The detection probability is therefore

$$P_d = \int_\eta^\infty \exp\left[-\frac{(x - 2E/N_0)^2}{2(2E/N_0)}\right] \frac{dx}{(2\pi(2E/N_0))^{1/2}}$$

$$\approx \frac{1}{2}\left[1 + \operatorname{erf}\left(\sqrt{\frac{E}{N_0}} - \gamma\right)\right] \tag{5.52}$$

which, as noted previously, depends only on signal-to-noise ratio E/N_0, and of course P_{fa} through the parameter γ. However, for any finite value of γ, P_d can be made arbitrarily close to unity by making E/N_0 arbitrarily large. The prescription is therefore

1. Determine the numerical parameter γ from the desired false alarm probability using (5.51).
2. Determine the detection threshold from $\eta = 2\gamma\sqrt{E/N_0}$.
3. For this detection threshold the detection probability will be given by (5.52). Numerical examples will be presented in Chapter 6, where coherent and non-coherent detection are evaluated and compared. Note that in (5.52) the dimensionless variance is equal to the mean and we have a one-parameter distribution, to which the discussion of Section 2.2 applies. Dependence of P_d on $\sqrt{E/N_0}$ may be understood as follows. The operating characteristics of the receiver improve as the overlap between the distributions $P(y|H_0)$ and $P(y|H_1)$ becomes smaller. Since the separation between the distributions is $2E/N_0$ and their standard deviation is $\sqrt{2E/N_0}$, the overlap becomes smaller as the ratio of these quantities, or equivalently $\sqrt{E/N_0}$, increases.

The properties of the matched filter shall be discussed in some detail shortly, after dealing with shot noise. Before preceding with this however it is worth noting an important consideration. In obtaining (5.42) it is tacitly assumed that although the sampling interval Δt becomes vanishingly small, successive noise samples separated by Δt seconds remain independent because B is arbitrarily large. Thus (5.42) is valid only if all preceding stages in the processing path are comparatively wide-band in comparison with the filter bandwidth $\sim 1/T$. If this is not the case then (5.42) does not follow from (5.39) because successive noise samples will not be independent; similar considerations arise in the analysis of phase-locked loops. This issue will arise again in Chapter 7. It is also important to keep in mind that the use of correlation detection, and therefore matched filtering, to implement the likelihood ratio test arises as a result of the quadratic exponent in (5.37) which of course is characteristic of the Gaussian distribution. It will be seen for the case of shot noise that implementation of the LRT does not involve correlation of the observable with a stored replica of the signal.

5.4.2 The Shot Noise Channel

Referring to Section 5.3.2, the photoelectric current which is observed at any given instant consists of a superposition of the responses to all the elementary current impulses produced by photon arrivals during the previous $\tau = 1/2B$ seconds. Also, successive observations at t_i and t_j are independent if $(t_i - t_j) > \tau$. Therefore if the signal current is of duration T, then there are $T/\tau = 2BT$ independent observations of the random process $k(t)$ and the LRT—the extension of (5.26) to this case—is

$$\prod_{j=1}^{2BT} \frac{P(k(t_j)|H_1)}{P(k(t_j)|H_0)} = \prod_{j=1}^{2BT} \left(1 + \frac{i_s(t_j)}{i_d(t_j)}\right)^{k(t_j)} e^{-i_s(t_j)\tau/q} > \eta \tag{5.53}$$

or equivalently

$$\sum_{j=1}^{2BT} k(t_j) \log\left(1 + \frac{k_s(t_j)}{k_d(t_j)}\right) > \log \eta + \sum_{j=1}^{2BT} k_s(t_j) = \log \eta + K_s \tag{5.54}$$

where k_s and k_d are the counts in an interval τ due to signal and dark current alone respectively.

To obtain a continuous-time representation we write

$$\sum_{j=1}^{2BT} k(t_j) \log\left[1 + \frac{k_s(t_j)}{k_d(t_j)}\right] = \sum_{j=1}^{2BT} k(t_j) \log\left[1 + \frac{k_s(t_j)}{k_d(t_j)}\right] \frac{\Delta t}{\Delta t}$$

$$\approx 2B \int_0^T k(t) \log\left[1 + \frac{k_s(t)}{k_d(t)}\right] dt$$

and (5.49) becomes

$$2B \int_0^T k(t) \log\left[1 + \frac{k_s(t)}{k_d(t)}\right] dt > \log \eta + K_s \tag{5.55}$$

where $k(t)$ is the observed photoelectron count and K_s is the integrated total photoelectron count due to signal alone over the duration T, which is assumed known as is the received signal energy in (5.41). As before, the maximum-likelihood solution is obtained from the Bayes solution, (5.55) by setting $\eta = 1$.

For the Neyman–Pearson criterion it is necessary to deal with the integrated dark-current count K_d, also assumed known

$$K_d = \sum_{j=1}^{2BT} \frac{i_d(t_j)\tau}{q} = \sum_{j=1}^{2BT} k_d(t_j) \tag{5.56}$$

and the integrated signal-plus-dark-current count

$$\sum_{i=1}^{2BT} \frac{i_s(t_j) + i_d(t_j)}{2Bq} = K_d + K_s = \rho_T \tag{5.57}$$

The threshold K_0 is determined from

$$P_{fa} = 1 - \sum_{n=0}^{K_0-1} \frac{e^{-K_d} K_d^n}{n!} \tag{5.58}$$

yielding a detection probability

$$P_d = 1 - \sum_{n=0}^{K_0-1} \frac{e^{-(\rho_T)} (\rho_T)^n}{n!} \tag{5.59}$$

And of course it is the integrated count K

$$K = \int_0^T k(t)\, dt \tag{5.60}$$

a random variable, which is compared with K_0.

In implementing (5.55) there is an exact parallel to (5.42), with $\log\{1 + [i_s(t)/i_d(t)]\}$ substituted for $s(t)$ in Figure 5.5 and $k(t)$ replacing $y(t)$. Clearly, however, defining an impulse response to perform this operation would not yield a matched filter, as it does for Gaussian noise. We now consider some properties of the matched filter.

5.5 THE MATCHED FILTER—II

It has been shown in the preceding section that for Gaussian noise the matched filter provides the optimum operation on the observable, from the point of view of maximizing the probability of a correct decision regarding whether H_0 or H_1 is true or maximizing P_d for a given value of P_{fa}. In this, at no time did the question of maximizing the signal-to-noise ratio (SNR) arise. SNR however is of course an important consideration, which shall now be considered in connection with the operation of the matched filter. SNR can be defined in many ways. For purposes of simply detecting whether or not the signal is present imagine the appearance of the oscilloscope trace of the output of a filter whose input is noise of arbitrary statistics, having zero mean. The average level of the trace will also be zero, but the trace will fluctuate, with mean-square fluctuations equal to the variance of the noise, in this case the average noise power. If a signal is now introduced at the input the average level of the trace will deflect accordingly, the random fluctuations remaining the same. Clearly, it is the fluctuations due to noise which interfere with detecting the presence of the signal, as indicated by this deflection. If SNR is to be measure of our ability to detect the signal it therefore seems reasonable to define it as

$$\text{SNR} = \frac{|x_s(T)|^2}{\text{Average Noise Power}} \tag{5.61}$$

where $x_s(T)$ is the peak response due to signal only in (5.47).

Now by Parseval's theorem (3.79)

$$\int_0^T s^2(t)\, dt = \int_{-\infty}^{\infty} |S(f)|^2\, df = E \tag{5.62}$$

and we can therefore write

$$|x_s(T)|^2 = \left|\int_0^T s^2(t)\, dt\right|^2 = E\int_{-\infty}^{\infty} |S(f)|^2\, df \tag{5.63}$$

where E is the signal energy. For the noise, with two-sided power spectral density $N_0/2$, the output power spectral density is $|S(f)|^2 N_0/2$ and therefore

$$\text{SNR} = \frac{E\int_{-\infty}^{\infty} |S(f)|^2\, df}{\frac{N_0}{2}\int_{-\infty}^{\infty} |S(f)|^2\, df} = \frac{2E}{N_0} \tag{5.64}$$

where in the denominator of (5.64), (3.12) has been used.

Equation (5.64) expresses the important result that SNR as defined in (5.61) at the output of a matched filter is the ratio of two fundamental parameters, E and $N_0/2$, and is independent of the particular shape of the signal waveform or any other signal parameter, including its bandwidth. This of course arises from the fact that the matched filtering operation performs the correlation of the signal with itself.

Now consider the optimum operation on the observable from the point of view of maximizing SNR for a signal in additive noise. We wish to find that $h(t)$ which maximizes

$$\text{SNR} = \frac{|\int_{-\infty}^{\infty} h(\tau)s(T-\tau)\, d\tau|^2}{\int_{-\infty}^{\infty} |H(f)|^2 W_N(f)\, df} = \frac{|\int_{-\infty}^{\infty} H(f)S(f)e^{i2\pi fT}\, df|^2}{\int_{-\infty}^{\infty} |H(f)|^2 W_N(f)\, df} \tag{5.65}$$

where $H(f)$ and $S(f)$ are the Fourier transforms of $h(t)$ and $s(t)$ and $W_N(f)$ is the power spectral density of the input noise. Recall the Schwarz inequality

$$\left|\int f(t)g(t)\, dt\right|^2 \le \int |f(t)|^2\, dt \int |g(t)|^2\, dt \tag{5.66}$$

with equality iff $f(t) = kg^*(t)$ where k is a constant. Therefore, by multiplying and dividing the integrand of the numerator of (5.65) by $\sqrt{W_N(f)}$ and using (5.66)

$$\text{SNR} = \frac{\left| \int_{-\infty}^{\infty} H(f)\sqrt{W_N(f)} \dfrac{S(f)}{\sqrt{W_N(f)}} e^{i2\pi fT} \, df \right|^2}{\int_{-\infty}^{\infty} |H(f)|^2 W_N(f) \, df}$$

$$\leq \frac{\int_{-\infty}^{\infty} |H(f)|^2 W_N(f) \, df \int_{-\infty}^{\infty} |S(f)|^2 / W_N(f) \, df}{\int_{-\infty}^{\infty} |H(f)^2 W_N(f) \, df}$$

$$= \int_{-\infty}^{\infty} (|S(f)|^2 / W_N(f)) \, df \tag{5.67}$$

and the maximum is achieved iff

$$H(f) = k \frac{S^*(f)e^{-i2\pi fT}}{W_N(f)} \tag{5.68}$$

where * is not needed in the denominator because $W_N(f)$ is always real (and non-negative). Therefore, if the noise is white—that is if $W_N(f)$ is a constant k over the band of $H(f)$—then

$$h(t) = s(T - t) \tag{5.69}$$

which is the impulse response of a filter matched to $s(t)$ as derived previously. Note that it is not necessary that the noise be Gaussian for (5.64) and (5.69) to hold, but only that its spectrum be flat over the frequency band of interest. Thus, to summarize, the matched filter maximizes SNR over all probability densities, provided the power spectral density is a constant. In the event that the noise PSD is not flat (colored noise), the matched impulse response corresponds to the modified signal spectrum $S^*(f)e^{-i2\pi fT}/W_N(f)$ rather than simply $S^*(f)e^{-i2\pi fT}$. In this way operations equivalent to matched filtering, for colored noise, can be derived [19], which however shall not be considered further here.

EXERCISES FOR CHAPTER 5

5.1 For the Bayes and Maximum Likelihood criteria, η determines P_{fa} and P_{d}. Show for these two cases that these quantities are, respectively, for Gaussian noise as discussed in Section 5.4.1.

Bayes

$$P_{\text{fa}} = \frac{1}{2}\left(1 - \text{erf}\left(\frac{\sigma \log \eta}{s\sqrt{2}} + \frac{s}{2\sigma\sqrt{2}}\right)\right)$$

$$P_{\text{d}} = \frac{1}{2}\left(1 + \text{erf}\left(\frac{s}{2\sigma\sqrt{2}} - \frac{\sigma \log \eta}{2\sqrt{2}}\right)\right)$$

Maximum Likelihood

$$P_{\text{fa}} = \frac{1}{2}\left(1 - \text{erf}\left(\frac{s}{2\sigma\sqrt{2}}\right)\right)$$

$$P_{\text{d}} = \frac{1}{2}\left(1 + \text{erf}\left(\frac{s}{2\sigma\sqrt{2}}\right)\right)$$

Show that for Maximum Likelihood

$$P_{\text{d}} = 1 - P_{\text{fa}}$$

5.2 Plot (5.25) for $\lambda T = 0$, 0.01, 0.1, 1, 2, 5, 10, 100,000. For $k_{\text{d}} = 1$ and $\rho = 10$, find P_{d} for $P_{\text{fa}} = 0.001$ [see (5.34) and (5.35)].

5.3 Prove that the functions $\sin \pi(x-n)/\pi(x-n)$ are orthonormal on n over the interval $-\infty \le x \le \infty$. That is

$$\int_{-\infty}^{\infty} \frac{\sin \pi(x-n)}{\pi(x-n)} \frac{\sin \pi(x-m)}{\pi(x-m)}\, dx = \delta_{n,m}$$

where $\delta_{n,m}$ is the Kronicker delta:

$$\delta_{n,m} = \begin{cases} 1 & n = m \\ 0 & n \ne m \end{cases}$$

5.4 Let $h(t) = t[u(t) - u(t-T)]$ and $y(t) = u(t)e^{-\lambda t}$, where $u(t)$ is the unit step function

$$u(t) = \begin{cases} 1, & t \ge 0 \\ 0, & \text{otherwise} \end{cases}$$

Show that $x(t)$ in 4.44 is equal to

$$\frac{1}{\alpha}\left(t - \frac{1}{\alpha}\right) + \frac{e^{-\alpha t}}{\alpha^2}$$

for $0 \le t \le T$ and is equal to

$$\frac{e^{-\alpha(t-T)}}{\alpha}\left(T - \frac{1}{\alpha}\right) + \frac{e^{-\alpha t}}{\alpha^2}$$

for $t > T$.

5.5 For $y(t)$ and $h(t)$ both equal to zero for $t<0$ show that the two following forms of $x(t)$,

$$x(t) = \int_0^t h(\tau)y(t-\tau)\, d\tau$$

and

$$x(t) = \int_{-\infty}^{\infty} h(\tau)y(t-\tau)\, d\tau$$

are identical. We can also write

$$x(t) = \int_0^t y(\tau)h(t-\tau)\, d\tau$$

What determines the limits 0 and t in this case?

5.6 For a certain random variable x there are two hypotheses: H_0, that x is normal (μ_0, σ) and H_1 that x is normal (μ_1, σ). Write down the LRT for choosing between H_0 and H_1 using the Bayes criterion, and the necessary condition on the observable x for selection of H_1. What is the condition for the Maximum Likelihood criterion?

5.7 Apply the Neyman–Pearson criterion to Exercise 5.6. Let the false alarm probability be α. Write down an expression for the detection probability in terms of μ_0 and μ_1, identifying all parameters and showing how they would be evaluated.

5.8 Suppose the filter characteristics of Figure 5.4 were as shown. How does this change things in Section (5.4.1)? Explain.

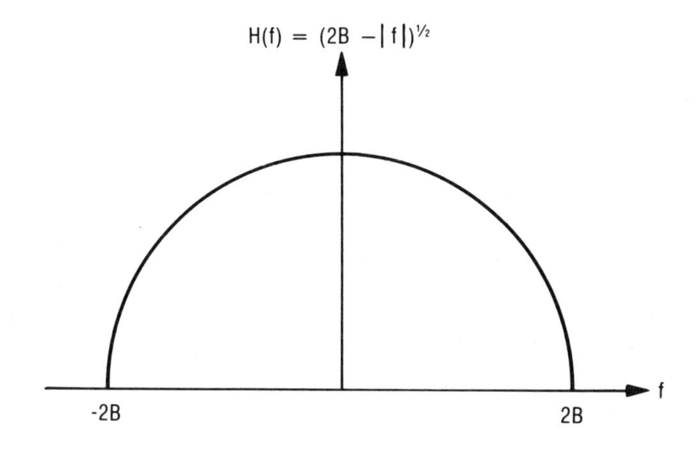

5.9 For a signal $s(t)$ as shown, sketch the matched-filter impulse response and write down an expression for maximum output in terms of the signal energy E.

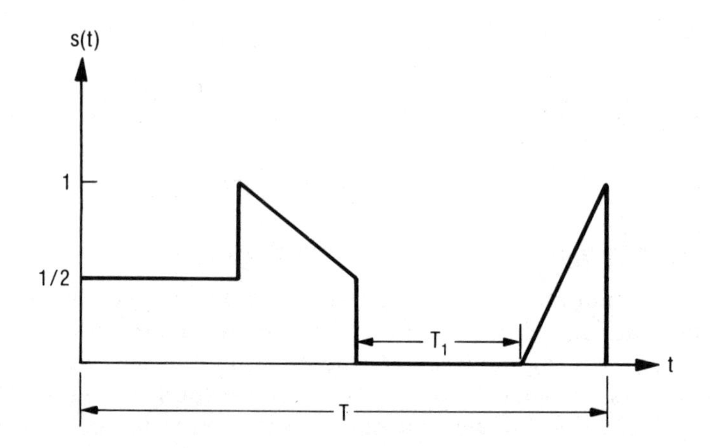

5.10 The Rayleigh and Rice distributions were introduced in Chapter 2. If observations are made at the output of an envelope detector, write down the LRT for a Bayes decision scheme to choose between the hypotheses: H_0, that no signal present, noise power $= \sigma^2$; H_1 that signal of the form $A \cos 2\pi f_0 t$ is present.

5.11 A coin is to be tested to see if it is fair or not. The two hypotheses are H_0: $P(\text{head}) = P(\text{tail}) = \frac{1}{2}$, H_1: $P(\text{head}) = 0.75$, $P(\text{tail}) = 0.25$. The decision is to be based on 5 tosses. Write down the decision rule based on the maximum likelihood criterion and determine the decision threshold.

If the coin is to be chosen from a barrel containing 100 coins, 80 of which are biased and 20 of which are not, write down the LRT which maximizes the probability of a correct decision and determine the decision threshold.

5.12 **a.** Suppose a signal of interest is as shown. Let the noise be mean-zero, white Gaussian with variance σ^2, and assume that noise observations separated in time by $1/2B$ seconds are independent. For an observation over a time T, write the LRT for deciding between the hypotheses H_1 that the signal is present and H_0 that it is not. Use the Bayes criterion.

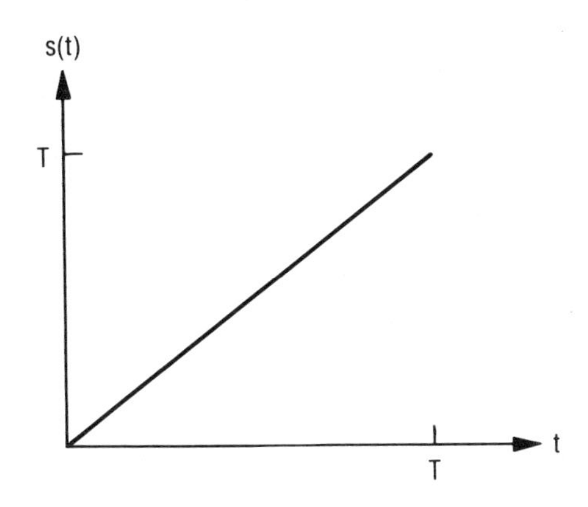

5.12 **b.** Repeat for the following signal.

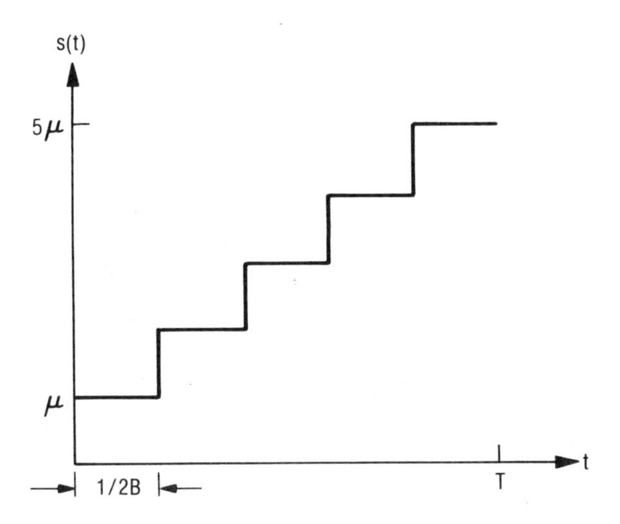

5.13 Consider the signal shown. Of the following impulse responses, which are causal, which are matched to $s(t)$? Repeat for the following cases and sketch the responses $x(t) = s(t) * h(t)$.

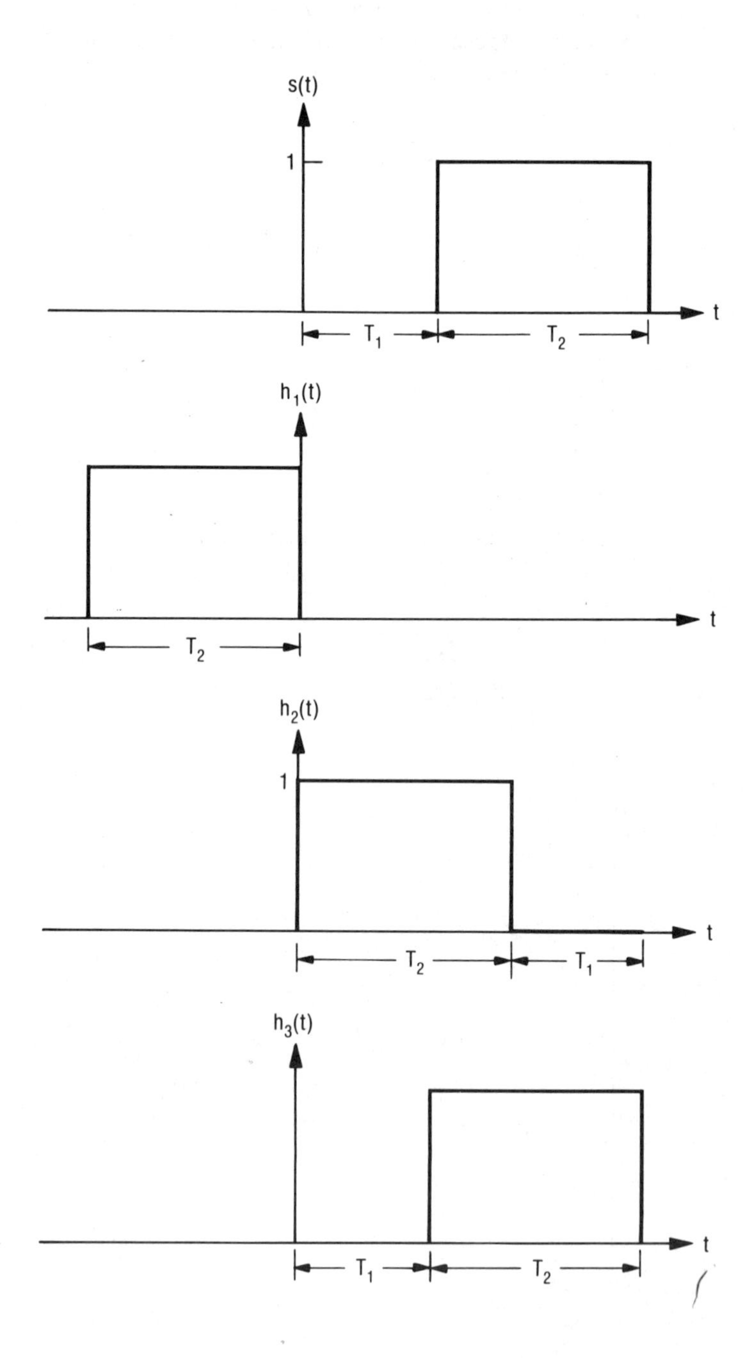

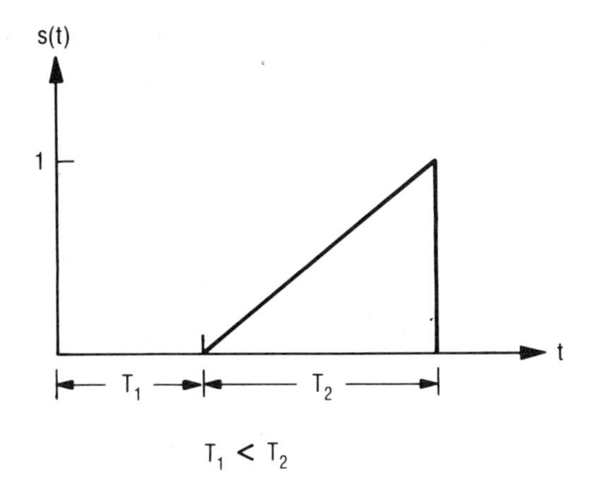
s(t)
1
t
T1
T2
T1 < T2

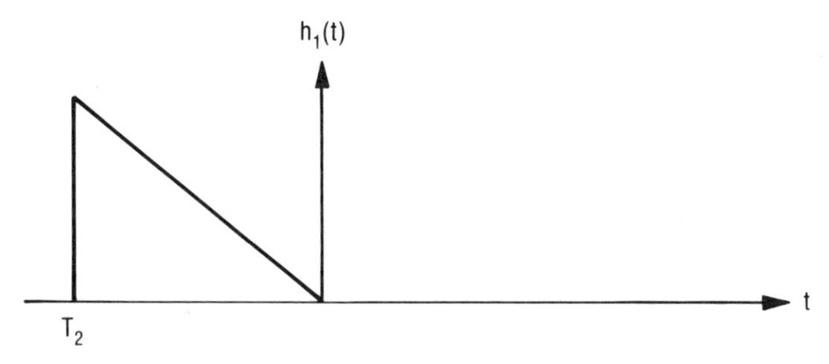
h1(t)
t
T2

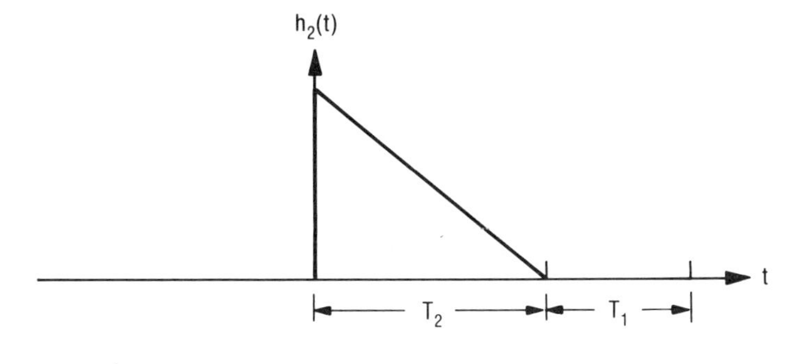
h2(t)
t
T2
T1

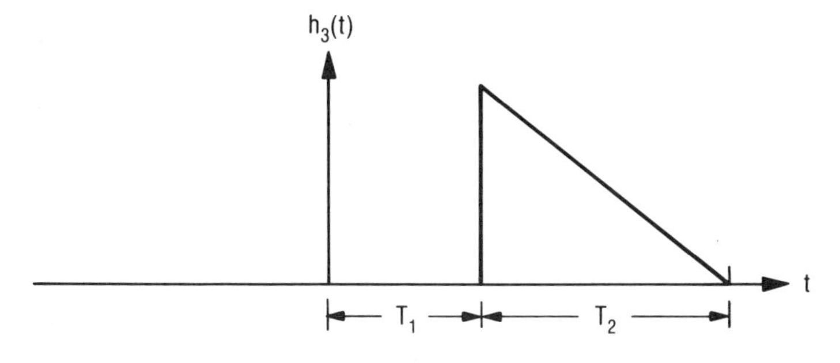
h3(t)
t
T1
T2

5.14 A target has an a priori probability of 3/4 of being present—that is $P(H_0) = 0.25$, $P(H_1) = 0.75$. The observation is to be made at the output of a matched filter, with $E/N_0 = 20$. For a required value of $P_{\text{fa}} = 0.00001$, find the threshold and P_d under the Neyman–Pearson criterion. Then, using the threshold given by (5.42) find P_{fa} and P_{d} under the Bayes and maximum likelihood criteria. Compare with the Neyman–Pearson results and comment on the differences in terms of the somewhat larger values of P_{d} at the expense of orders of magnitude larger values of P_{fa}.

6

COHERENT AND NONCOHERENT DETECTION AND PROCESSING

The question concerning the relative advantages of coherent and noncoherent detection and processing arises in all sensing systems, and also in communication systems, and is concerned with evaluating the benefits associated with preserving phase information in the signal in the processing. In Chapter 5 it was shown that matched filtering involves a correlation of the received signal with an exact replica of itself, which therefore includes its phase. Hence matched filtering is synonymous with coherent detection, in the sense that the phase of the signal is assumed to be known exactly. There is however another form of coherent detection and processing in which the transmitted waveform consists of a train of pulses whose responses are summed, or "integrated" in a manner such that the phase relationship between successive pulses is maintained, but the absolute value of the phase may or may not be known exactly. This is referred to as coherent integration. If on the other hand the summation is performed without regard to the phase relationship between pulses the process is called non-coherent integration.

In this chapter, we first consider noncoherent detection of a single pulse and compare the results with coherent detection for which, as noted, the results have already been obtained in Chapter 5. Then the improvement in SNR yielded by coherent and non-coherent integration of a train of pulses shall be calculated and compared. Finally the performance of coherent and noncoherent integration in terms of detection and false alarm probabilities shall be analyzed and compared. A summary of the results of this chapter is presented in Section 6.5.

As is discussed in Chapter 1, the processing of sensor data may include translation of the carrier frequency down to a more manageable IF. This step however is irrelevant to the results which follow, and to simplify the presentation is therefore omitted.

6.1 IDEAL NONCOHERENT DETECTION OF A SINGLE PULSE

Consider a situation in which the signal of interest is a rectangular carrier pulse of duration $T^{\dagger}$ of the form $AP_T(t)\cos(2\pi f_0 t + \theta)$ where

$$P_T(t) = \begin{cases} 1, & 0 \le t \le T \\ 0, & \text{otherwise} \end{cases}$$

The two hypotheses are

$$\begin{aligned} H_1&: \quad y(t) = s(t) + n(t) = AP_T(t)\cos(2\pi f_0 t + \theta) + n(t) \\ H_0&: \quad y(t) = n(t) \end{aligned} \tag{6.1}$$

where $n(t)$ is a realization of mean-zero, white Gaussian noise with two-sided spectral density $N_0/2$. In this case, in parallel with the discussion in Section 5.4.1, it is assumed that observations of $y(t)$ are made at the output of a hypothetical rectangular bandpass filter of bandwidth B centered at the carrier frequency f_0, as illustrated in Figure 6.1.

The bandwidth B must be sufficiently large to pass essentially all the signal energy, which will be the case if $B \ge 1/T$. In this case, by the carrier-sampling theorem, with sampling at the Nyquist rate there are BT *complex* samples in a time T and, in a manner essentially identical to that used to obtain (5.42) from (5.39) (also see Exercise 6.1), the likelihood ratio in this case is

$$\begin{aligned} &\frac{\exp[-(1/2\sigma^2\, \Sigma_{i=1}^{BT}\, (y(t_i) - s(t_i))^2]}{\exp[-(1/2\sigma^2)\, \Sigma_{i=1}^{BT}\, y^2(t_i)]} \\ &\quad = \frac{\exp[-(2B/2\sigma^2)\int_0^T (y(t) - A\cos(2\pi f_0 t + \theta))^2\, dt]}{\exp[-(2B/2\sigma^2)\int_0^T y^2(t)\, dt]} \end{aligned} \tag{6.2}$$

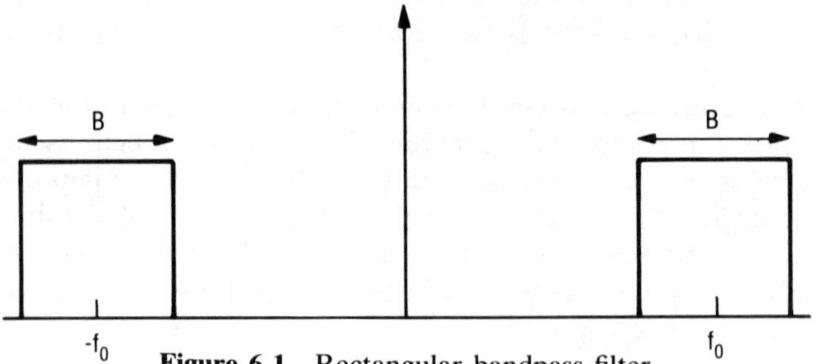

Figure 6.1 Rectangular bandpass filter.

† A generalized treatment of this subject using arbitrary complex signals is presented in Chapter 10.

As before, because the signal is of finite duration, the approximation in passing from the discrete to the continuous representation improves in the limit as B is allowed to become very large.

In the non-coherent case θ in (6.2) is unknown, and the likelihood ratio will therefore be of the form $P(y|H_1, \theta)/P(y|H_0)$, which will have a different unknown value for each value of the parameter θ. Therefore, since no auxiliary information about θ is available, it is reasonable to assume θ to be uniformly distributed over $(0, 2\pi)$ and we deal with an averaged likelihood ratio of the form

$$\int_0^{2\pi} \frac{P(y|H_1, \theta)P(\theta)}{P(y|H_0)}\, d\theta$$
$$= \frac{1}{2\pi} \int_0^{2\pi} \left\{ \exp\left[\frac{2B}{\sigma^2} \int_0^T y(t) A \cos(2\pi f_0 t + \theta)\, dt - \frac{2B}{2\sigma^2} \int_0^T A^2 \cos^2(2\pi f_0 t + \theta)\, dt \right] \right\} d\theta \tag{6.3}$$

The integral in the exponent involving $\cos^2(2\pi f_0 t + \theta)$ is

$$\int_0^T \cos^2(2\pi f_0 t + \theta)\, dt = \frac{T}{2}\left[1 + \frac{1}{T}\int_0^T \cos(4\pi f_0 t + 2\theta)\, dt\right]$$
$$= \frac{T}{2}\left[1 + \frac{1}{2\pi f_0 T} \sin 2\pi f_0 T \cos(2\pi f_0 T + \theta)\right] \approx \frac{T}{2} \tag{6.4}$$

where the approximation is very good for $f_0 T$, the number of cycles per pulse, greater than about 3 or 4, which will be assumed to be the case. Therefore, since $\sigma^2 = (N_0/2)(B + B) = N_0 B$ the likelihood ratio is

$$\frac{\exp(-A^2T/2N_0)}{2\pi} \int_0^{2\pi} \left\{ \exp\left[-\frac{2A}{N_0} \int_0^T y(t) \cos(2\pi f_0 t + \theta)\, dt \right] \right\} d\theta \tag{6.5}$$

The exponent in the integrand can be written as

$$-\frac{2A}{N_0} \int_0^T y(t) \cos(2\pi f_0 t + \theta)\, dt$$
$$= -\frac{2A}{N_0} I \cos\theta + \frac{2A}{N_0} Q \sin\theta$$
$$I = \int_0^T y(t) \cos 2\pi f_0 t\, dt\,, \quad Q = \int_0^T y(t) \sin 2\pi f_0 t\, dt \tag{6.6}$$

and (6.5) becomes, using (3.52)

$$\exp\left(-\frac{A^2T}{2N_0}\right)\int_0^{2\pi}\exp\left[-\frac{2A}{N_0}(I\cos\theta - Q\sin\theta)\right]\frac{d\theta}{2\pi}$$

$$= \exp\left(-\frac{A^2T}{2N_0}\right)I_0\left(\frac{2Az}{N_0}\right) \tag{6.7}$$

where $z = \sqrt{I^2 + Q^2}$ and I_0 is the modified Bessel function of order zero.† The likelihood Ratio Test (LRT) is therefore: choose H_1 if

$$I_0\left(\frac{2Az}{N_0}\right) \geq e^{E/N_0}\eta \tag{6.8}$$

and choose the null hypothesis H_0 otherwise; in (6.8) the exponent has been written in terms of the signal energy $E = A^2T/2$. Thus the modified Bessel function I_0 is the optimum non-coherent detection characteristic.

In what follows only the Neyman–Pearson criterion shall be applied. Since $I_0(x)$ is a monotonically increasing function of x, the LRT (6.8) can be expressed in an equivalent form as: choose the alternative H_1 if either

$$z = \sqrt{I^2 + Q^2} \geq \eta \quad \text{or} \quad z^2 = I^2 + Q^2 \geq \eta^2 \tag{6.9}$$

where the threshold η (not necessarily the same as that in (6.8)) is determined by the specified false alarm probability. In what follows we choose the former case and deal here with z and η. In Chapter 10, which deals with this subject using generalized complex signals, identical results are obtained using square-law detection which demonstrates the exact equivalence of envelope and square-law detection in this case.

Before dealing with the determination of η for the non-coherent case, let us consider the generation of the statistic z. This can be accomplished by the system illustrated in Figure 6.2 which is customarily referred to as a quadrature receiver, in which the paths employing multiplication by $\cos 2\pi f_0 t$ and $\sin 2\pi f_0 t$ are the in-phase and quadrature channels, yielding I and Q respectively.

Figure 6.2 is reminiscent of the correlation detector of Figure 5.4. That is, as discussed in connection with that figure, the multiplication-and-integration processes in the in-phase and quadrature channels in Figure 6.2 are equivalent to filters $h(t)$ matched to $P_T(t)\cos 2\pi f_0 t$ and $P_T(t)\sin 2\pi f_0 t$ respectively. Furthermore, the squaring adding and square-root operations which follow amount to envelope detection. The statistic z can therefore be generated by the system illustrated in Figure 6.3. To show this, note that for input $y(t)$ the output at any arbitrary time t is

† There should be no confusion between the notation I_0, the zero-order modified Bessel function, and I or I_i which denotes the in-phase channel term.

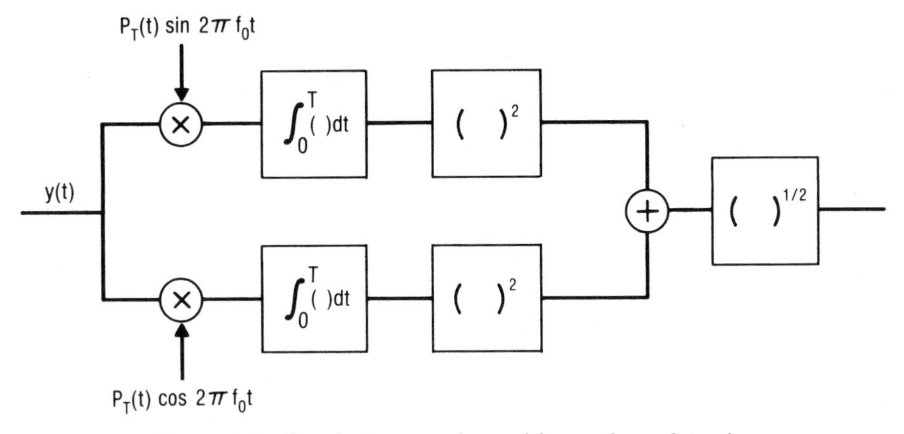

Figure 6.2 Quadrature receiver with envelope detection.

$$
\begin{aligned}
\int_0^t y(\tau)h(t-\tau)\,dt &= \int_0^t y(\tau)\cos 2\pi f_0(T-t+\tau)\,d\tau \\
&= \cos 2\pi f_0(T-t)\int_0^t y(\tau)\cos 2\pi f_0\tau\,d\tau \\
&\quad - \sin 2\pi f_0(T-t)\int_0^t y(\tau)\sin 2\pi f_0\tau\,d\tau
\end{aligned}
$$

and the envelope is

$$
\left[\left(\int_0^t y(\tau)\cos 2\pi f_0\tau\,d\tau\right)^2 + \left(\int_0^t y(\tau)\sin 2\pi f_0\tau\,d\tau\right)^2\right]^{1/2}
$$

which at the time of correlation, $t = T$, is equal to

$$
\left[\left(\int_0^T y(\tau)\cos 2\pi f_0\tau\,d\tau\right)^2 + \left(\int_0^T y(\tau)\sin 2\pi f_0\tau\,d\tau\right)^2\right]^{1/2} = \sqrt{I^2+Q^2} = z \tag{6.10}
$$

As is discussed in Chapter 1, the envelope detector is also referred to as a linear detector, which, as illustrated in Figure 6.4, refers to that property of the detector whereby the output is proportional to the input when the input is positive. The operation of the detector however is of course highly

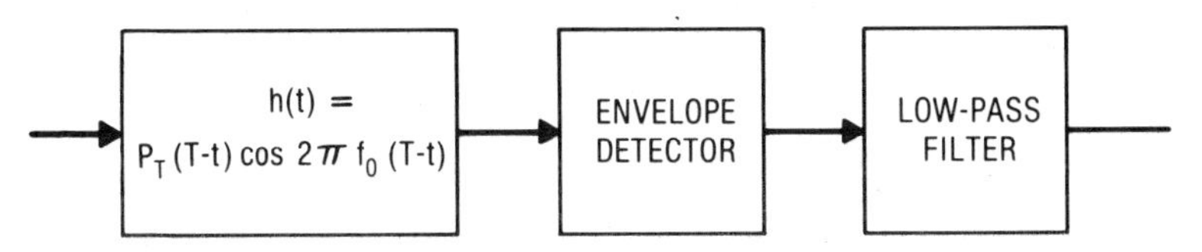

Figure 6.3 Equivalent of quadrature-receiver.

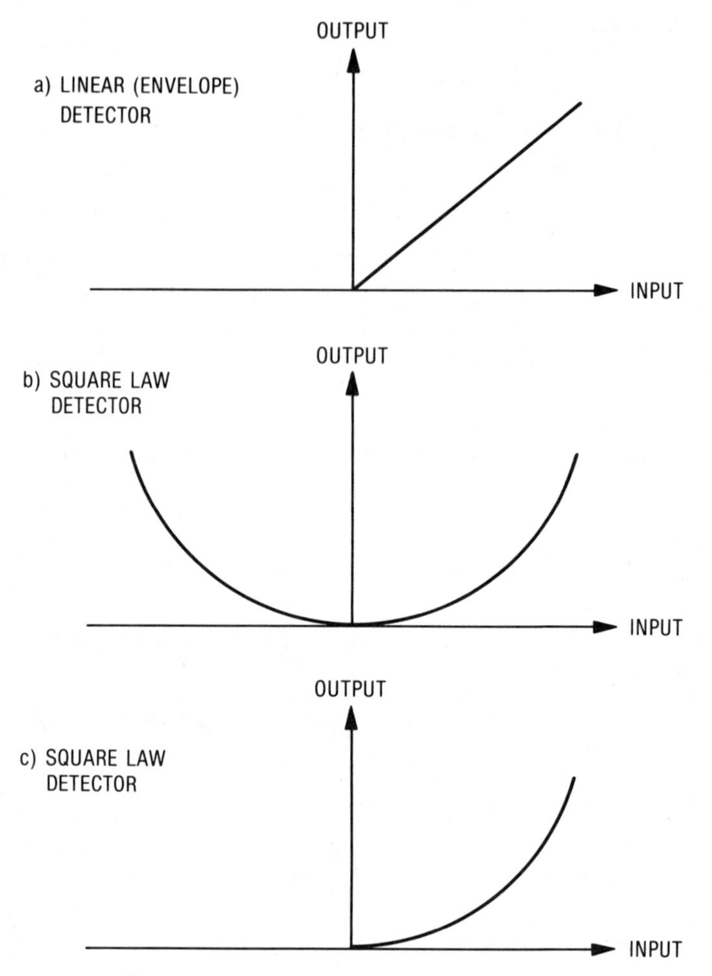

Figure 6.4 Various detector characteristics.

non-linear, as a result of which the output consists of a DC term equal to the envelope $\sqrt{I^2 + Q^2}$, the carrier term, plus an infinite number of harmonics of the input at $2f_0$, $3f_0$, etc. It is for this reason that, as discussed in Chapter 1, the output of the envelope detector must be passed through a low-pass filter or a video amplifier, which will eliminate the carrier term and the unwanted harmonics. Similar comments apply to square-law detectors which generate z^2, whose characteristics are shown in Figure 6.4c and d.

Now the transfer function of the filter in Figure 6.3 is

$$H(f) = \int_0^T \cos 2\pi f_0(T - t)e^{-i2\pi ft}\, dt$$

$$= \frac{e^{-i\pi(f-f_0)T}}{2} \frac{\sin \pi(f - f_0)T}{\pi(f - f_0)} + \frac{e^{-i\pi(f+f_0)T}}{2} \frac{\sin \pi(f + f_0)T}{\pi(f + f_0)} \quad (6.11)$$

which is essentially that of a bandpass filter of bandwidth $\sim 1/T$ with the indicated phase characteristic, which of course is irrelevant here since the filtering is followed by envelope detection. Hence, for non-coherent detection the filtering operation is in practice implemented by a bandpass filter, with essentially arbitrary phase characteristics, whose bandwidth is nominally equal to that of the signal. As is discussed in [20], the purpose of bandwidth matching is to maximize SNR. Suppose the bandwidth is very narrow. If it is gradually widened, both the signal and the noise energy at the output will increase, but the signal energy will increase faster because it builds up coherently over its duration whereas the noise contributions at successive time instants are independent. Thus SNR increases. When the bandwidth is nominally equal to that of the signal however, further widening leaves the signal energy at the output unchanged and serves only to increase the noise, and SNR decreases. Hence the optimum is achieved when the filter bandwidth equals the signal bandwidth.

The filter bandwidth in this case is therefore matched to the signal bandwidth. If one wishes to be more careful the actual spectral shape of the filter can be matched to that of the signal. However, the matching is in amplitude only, and we do not actually have a matched filter since, as is discussed in connection with Figure 5.5, true matched filtering is equivalent to having available an *exact* replica of the signal, which includes the phase.

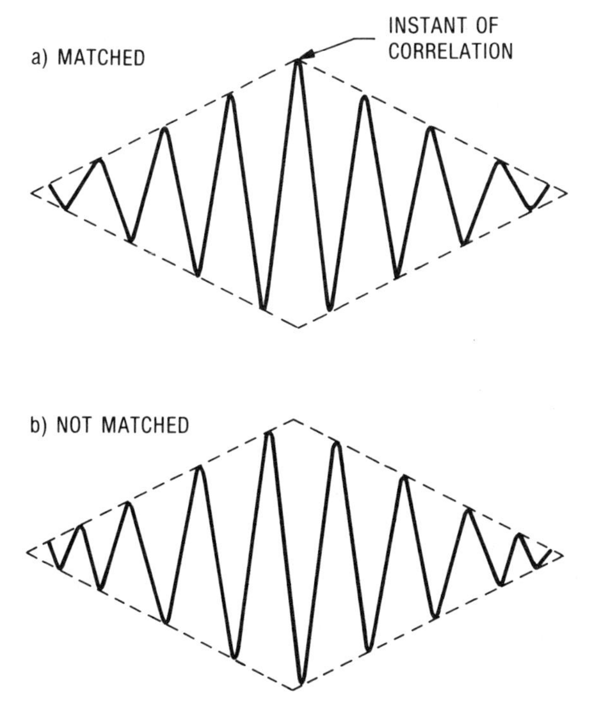

Figure 6.5 Outputs of matched and unmatched filters.

By the same token therefore, matched filtering is exactly equivalent to coherent detection and as noted, the results of Chapter 5 apply; one might denote the operation in this section as incoherent matched filtering [21]. To illustrate the difference between the two cases consider the outputs of: (a) a filter matched to a signal $P_T(t)A\cos(2\pi f_0 t + \theta)$ and (b) that of a filter whose impulse response is $P_T(t)A\sin 2\pi f_0(T-t)$, which are shown in Figure 6.5 (see Exercise 6.3). For the matched filter the instant of correlation (i.e. $t = T$) occurs simultaneously with a crest of the carrier yielding the maximum peak response. In the non-coherent case this will generally not be the case and the peak response may be smaller.

6.2 COMPARISON OF COHERENT AND NONCOHERENT DETECTION OF A SINGLE PULSE

We are now in a position to compare quantitatively the effectiveness of coherent and noncoherent detection of a single pulse. Referring to (6.6), the in-phase and quadrature components of the output of the bandpass filter in Figure 6.3 at $t = T$ are

$$I = \int_0^T y(t)\cos 2\pi f_0 t\, dt = \int_0^T A\cos(2\pi f_0 t + \theta)\cos 2\pi f_0 t\, dt$$

$$+ \int_0^T n(t)\cos 2\pi f_0 t\, dt = \frac{AT}{2}\cos\theta + \int_0^T n(t)\cos 2\pi f_0 t\, dt$$

$$Q = \frac{AT\sin\theta}{2} + \int_0^T n(t)\sin 2\pi f_0 t\, dt \tag{6.12}$$

where the approximation of (6.4) has been used. The quantities

$$X = \int_0^T n(t)\cos 2\pi f_0 t\, dt = I - \frac{AT\cos\theta}{2} \tag{6.13}$$

$$Y = \int_0^T n(t)\sin 2\pi f_0 t\, dt = Q - \frac{AT\sin\theta}{2}$$

are each mean-zero Gaussian random variables with variances

$$E(X^2) = E\int_0^T\int_0^T n(s)n(t)\cos 2\pi f_0 t\cos 2\pi f_0 s\, dt\, ds$$

$$E(Y^2) = E\int_0^T\int_0^T n(s)n(t)\sin 2\pi f_0 t\cos 2\pi f_0 s\, dt\, ds$$

and since for white noise

$$E(n(s)n(t)) = \frac{N_0}{2}\,\delta(s-t)$$

then to a very good approximation, again using (6.4)

$$E(X^2) = \frac{N_0}{2}\int_0^T \cos^2 2\pi f_0 t\, dt = \frac{N_0 T}{4} + \frac{N_0}{2}\int_0^T \cos 4\pi f_0 t\, dt$$
$$\approx \frac{N_0 T}{4} = \sigma_z^2 \tag{6.14}$$

with the same result for $E(Y^2)$. Therefore, since X and Y are independent Gaussian random variables their joint density is

$$P(X, Y)\, dX\, dY = \exp\left[-\frac{(X^2 + Y^2)}{2\sigma_z^2}\right]\frac{dX\, dY}{2\pi\sigma_z^2}$$
$$= \exp\left[-\frac{I^2 + Q^2 + (AT/2)^2 - AT(I\cos\theta + Q\sin\theta)}{2\sigma_z^2}\right]\frac{dX\, dY}{2\pi\sigma_z^2} \tag{6.15}$$

Now let $I = z\cos\phi$, $Q = z\sin\phi$, with $dX\, dY = dI\, dQ = z\, dz\, d\phi$. The exponent in (6.15) is then

$$-\frac{z^2 + (AT/2)^2 - zAT\cos(\theta - \phi)}{2\sigma_z^2} \tag{6.16}$$

and since $P(X, Y)\, dX\, dY = P(z, \phi)\, dz\, d\phi$, the probability density of the envelope z under H_1 is

$$P(z) = \int_0^{2\pi} P(z, \phi)\, d\phi = z\,\frac{\exp\{-[z^2 + (AT/2)^2]/2\sigma_z^2\}}{\sigma_z^2}$$
$$\times \frac{1}{2\pi}\int_0^{2\pi}\exp\left[-\frac{zAT}{2\sigma_z^2}\cos(\theta - \phi)\right]d\phi$$
$$= z\,\frac{\exp\{-[z^2 + (AT/2)^2]/2\sigma_z^2\}}{\sigma_z^2}\, I_0\left(\frac{zAT}{2\sigma_z^2}\right) \tag{6.17}$$

which is the Rice distribution that was introduced in Section 3.6, with a minor change in the argument (see Exercise 6.5).

Now under H_0, $A = 0$, and since $I_0(0) = 1$ (6.17) becomes under the null hypothesis the probability density for the envelope of noise alone, the Rayleigh distribution

$$P(z) = \frac{z}{\sigma_z^2}\exp\left(-\frac{z^2}{2\sigma_a^2}\right) \tag{6.18}$$

The threshold η in (6.8) is therefore determined from the false alarm probability P_{fa} using

$$P_{\mathrm{fa}} = \int_\eta^\infty z\,\frac{\exp(-z^2/2\sigma_z^2)}{\sigma_z^2}\, dz = e^{-\eta^2/2\sigma_z^2} \tag{6.19}$$

whence, using (6.14)

$$\eta = (-2\sigma_z^2 \log P_{\mathrm{fa}})^{1/2} = \left(-\frac{N_0 T}{2} \ln P_{\mathrm{fa}}\right)^{1/2} \tag{6.20}$$

and the detection probability P_{d} is

$$P_{\mathrm{d}} = \int_{\eta}^{\infty} z \, \frac{\exp[-(z^2 + (AT/2)^2)/2\sigma_z^2]}{\sigma_z^2} \, I_0\left(\frac{zAT}{2\sigma_z^2}\right) dz$$

This can be put into a more convenient dimensionless form with the change of variables $x = z/\sigma_z$ from which

$$P_{\mathrm{d}} = \int_{(-2\ln P_{\mathrm{fa}})^{1/2}}^{\infty} x \exp\left[-\frac{(x^2+\alpha^2)}{2}\right] I_0(x\alpha)\, dx \tag{6.21}$$

where $\alpha^2 = 2E/N_0$.

The expression in (6.21) is known as Marcum's Q function [22], and shows explicitly that detection probability depends only on P_{fa} and on the signal-to-noise ratio E/N_0. For coherent detection the results for the matched filter—(5.52) et seq.—are

$$P_{\mathrm{d}} = \frac{1}{2}\left[1 + \mathrm{erf}\left(\sqrt{\frac{E}{N_0}} - \gamma\right)\right] \tag{6.22}$$

where the numerical parameter γ is determined from P_{fa} through the relationship

$$P_{\mathrm{fa}} = \tfrac{1}{2}[1 - \mathrm{erf}(\gamma)] \tag{6.23}$$

and the detection threshold η is

$$\eta = 2\gamma\sqrt{\frac{E}{N_0}} \tag{6.24}$$

Note that, in addition to P_{fa}, the detection threshold depends on the received signal energy in the coherent case (6.23), but only on the noise and the integration time in the non-coherent case (6.20), which is somewhat more convenient.

A comparison of coherent and non-coherent detection using (6.21), (6.22) and (6.23) is presented in Figure 6.6 in terms of the value of P_{d} that can be achieved for a given value of P_{fa} as a function of signal-to-noise ratio E/N_0. It is seen that: (1) for small values of E/N_0—say $E_0 < 3$ dB—for any given value of P_{fa}, noncoherent detection requires from 2 to 3 dB more SNR than that required by coherent detection in order to achieve the same value of P_{d}; (2) for large values of E/N_0—say >10 dB—the difference in SNR

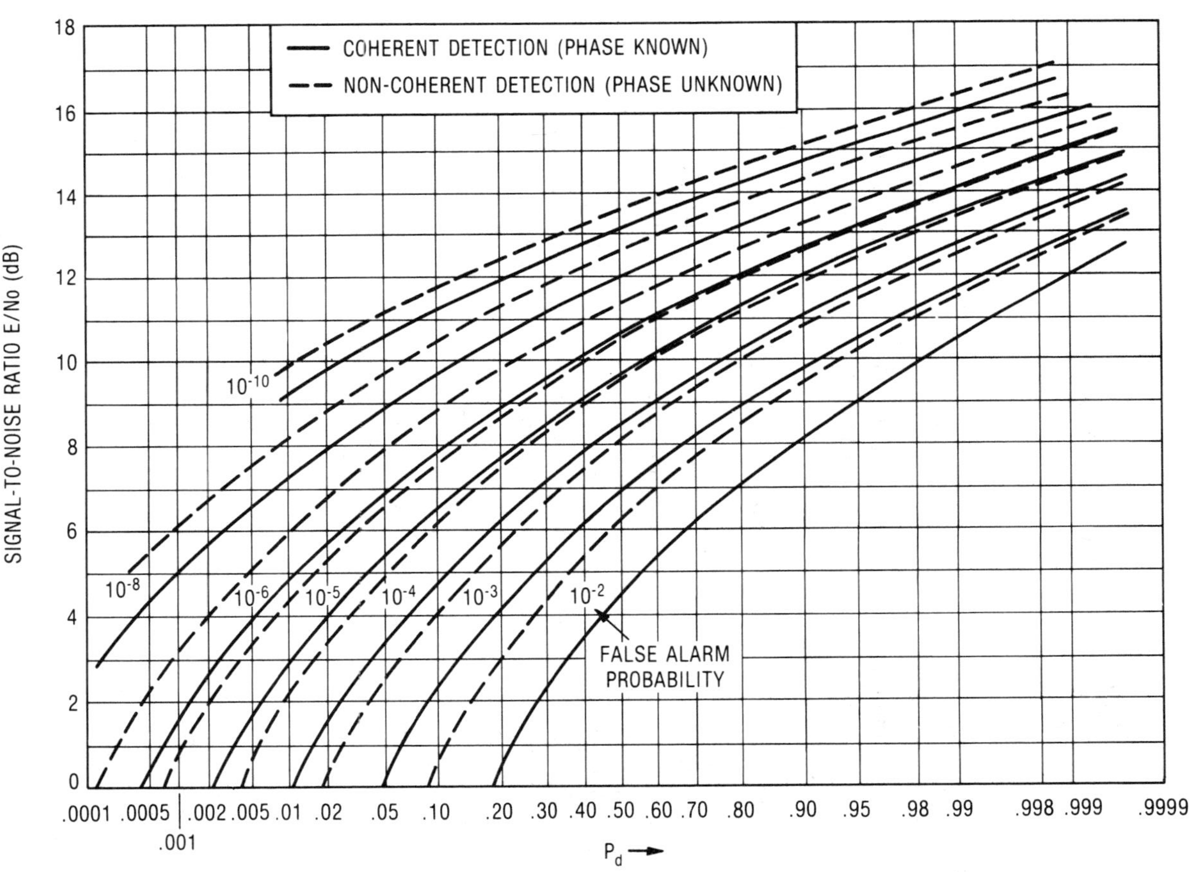

Figure 6.6 Detection probability for coherent and non-coherent detection of a single pulse (adapted from [21]).

required by the two schemes is less than ~1 dB, and it is clear that with further increase E/N_0 the difference eventually becomes negligible.

These results may be understood as follows. With a coherent matched filter the instantaneous sinusoidal amplitude of the filter output at some time t is

$$\left[\frac{At}{2} + \int_0^t n(\tau)\cos 2\pi f_0\tau\, d\tau\right]\cos 2\pi f_0(T-t)$$
$$-\left[\int_0^t n(\tau)\sin 2\pi f_0\tau\, d\tau\right]\sin 2\pi f_0(T-t) \tag{6.25}$$

where, again, (6.4) has been used and θ has been set equal to zero for convenience. In contrast with the non-coherent case (e.g. (6.12)), the signal component is contained entirely within the in-phase channel and, referring to (3.45), therefore contends with only the in-phase noise component; the quadrature channel need not come into play. Now the signal component at the matched filter output at the instant of correlation is $AT/2$ and from (6.14), $E(X^2) = E(Y^2) = N_0T/4$. Hence SNR for a matched filter is therefore $A^2T^2/N_0T = A^2T/N_0 = 2E/N_0$.

On the other hand, in the non-coherent case both the in-phase and quadrature noise components come into play, and in a similar calculation the noise term would be $N_0T/2$. This is clear from the Rayleigh distribution for the envelope of noise alone, for which $E(r^2) = 2\sigma^2$. Hence, if the signal amplitude A is not very large relative to σ, both noise components affect the random fluctuations in the envelope equally, and SNR is just E/N_0. However, if A is relatively large the fluctuations in the envelope of signal plus noise will be caused for the most part only by those noise components in phase with the signal and the signal-to-noise ratio is therefore effectively increased by 3 dB. To show this,† write the envelope of (6.24) as:

$$[(A+n_c)^2 + n_s^2]^{1/2} = A\left(1 + \frac{2n_c}{A} + \frac{n_c^2 + n_s^2}{A^2}\right)^{1/2}$$
$$\approx A + n_c + \frac{n_c^2 + n_s^2}{2A} \approx A + n_c \tag{6.26}$$

if A is large. Hence the fluctuations in the observable are carried primarily by n_c, and since $E(n_c^2) = \sigma^2 = N_0B = N_0/T$, then:

$$\frac{A^2}{\sigma^2} = \frac{A^2T}{N_0} = \frac{2E}{N_0} \tag{6.27}$$

From still another point of view, let us also recall (see Exercise 3.10) that if E/N_0 is large the Rice distribution can be approximated as a Gaussian,

† This can also be illustrated by a simple sketch based on Figure 3.9b.

with exponent $(r - A)^2/2\sigma^2$ and $A^2/\sigma^2 = 2E/N_0$. So to summarize, with a coherently matched filter, for which the input signal phase must be known exactly, an advantage over non-coherent detection is achieved when E/N_0 is small, which is equivalent to an improvement in E/N_0 by about a factor of two which is seen in Figure 6.6. On the other hand, if E/N_0 is large, say >10 dB, there is little difference between coherent and non-coherent matched filtering for a single pulse.

6.3 IMPROVEMENT IN SIGNAL-TO-NOISE RATIO BY COHERENT AND NONCOHERENT INTEGRATION

In the previous section coherent and noncoherent detection have been compared when the signal consists of a single pulse. In such cases the sensitivity of the sensor—that is, the maximum range at which a target can be reliably detected—is limited by the maximum available energy per pulse. The sensitivity can be significantly increased however, with no increase in peak power, by transmitting waveforms consisting of a large number of pulses and adding up their cumulative effect; this technique is also used to improve SNR in communication systems. In this case the average power $P_T T/T_p$, where T_p is the interpulse spacing, and P_T is the transmitted signal power per pulse, must be kept within the capability of the transmitter—the quantity T/T_p is known as the duty factor. As has been noted, there are two approaches to this type of signal processing, known as non-coherent integration and coherent integration, which will now be evaluated and compared in terms of the improvement in SNR that is achieved in the two cases.

6.3.1 Noncoherent Integration

The noncoherent case shall be considered first. Equation (6.7) gives the likelihood ratio for a single observation. It then follows at once that for M independent observations the likelihood ratio is

$$e^{-MA^2T/2N_0} \prod_{i=1}^{M} I_0\left(\frac{2Az_i}{N_0}\right) \tag{6.28}$$

where $z_i = \sqrt{I_i^2 + Q_i^2}$ is the ith output of the detector in Figure 6.3 and

$$I_i^2 = \left(\int_0^T y_i(\tau)\cos 2\pi f_0\tau \, d\tau\right)^2$$
$$Q_i^2 = \left(\int_0^T y_i(\tau)\sin 2\pi f_0\tau \, d\tau\right)^2 \tag{6.29}$$

where $y_i(t)$ is the ith observation, $i = 1, 2, \ldots, M$. The extension of the single-pulse LRT of (6.8) to a pulse-train waveform therefore becomes, after taking logarithms, declare H_1 if

$$\sum_{i=1}^{M} \ln I_0\left(\frac{2Az_i}{N_0}\right) \geq \frac{MA^2T}{2N_0} + \ln \eta \tag{6.30}$$

Now since the statistic consists of a sum of terms instead of a single term as in (6.8), the monoticity argument that was used there can no longer be applied and implementation of the I_0 characteristic, which as noted above is optimum for noncoherent detection, must be considered. It has been shown in Section 6.2 that the difference between coherent and noncoherent detection of a single pulse becomes negligible when SNR is large. In the interest of this comparison therefore let us consider the case when $2Az_i/N_0$ is small, and then make use of the approximation for small x

$$I_0(x) \approx 1 + x^2$$

Thus

$$\ln I_0\left(\frac{2Az_i}{N_0}\right) \approx \ln\left[1 + \left(\frac{2Az_i}{N_0}\right)^2\right] \approx \left(\frac{2Az_i}{N_0}\right)^2$$

and the I_0 characteristic is implemented by a square-law detector. Equation (6.30) therefore becomes

$$\sum_{i=1}^{M} z_i^2 \geq \frac{MTN_0}{8} + \frac{N_0^2}{4A^2} \ln \eta \tag{6.31}$$

The process of summing successive outputs after detection is illustrated in Figure 6.7. The integration is noncoherent because the phase information which contains the phase relationship between successive pulses is destroyed in the detection process.

For completeness, for large values of $2Az_i/N_0$, $I_0(2Az_i/N_0)$ is approximated by

$$I_0\left(\frac{2Az_i}{N_0}\right) \approx \frac{\exp(2Az_i/N_0)}{(4\pi Az_i/N_0)^{1/2}}$$

so that

$$\ln I_0\left(\frac{2Az_i}{N_0}\right) \approx \frac{2Az_i}{N_0} - \frac{1}{2}\ln\left(\frac{4\pi Az_i}{N_0}\right) \approx \frac{2Az_i}{N_0}$$

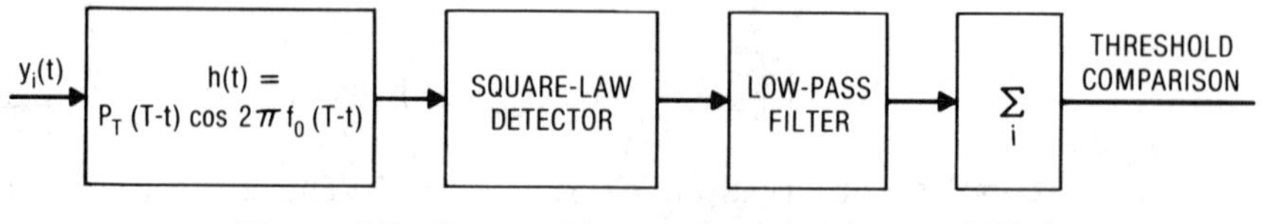

Figure 6.7 System for post-detection integration.

and (6.31) becomes, choose H_1 if

$$\sum_{i=1}^{M} z_i \geq \frac{MAT}{4} + \frac{N_0}{2A} \ln \eta \tag{6.32}$$

In this case the square-law detector in Figure 6.7 would be replaced by a linear, or envelope, detector. Thus, the rule for optimal implementation of the ideal Bessel function characteristic is to use linear detection for large SNR and square-law detection for small SNR. However there is little difference in performance for most cases of interest [21].

Continuing with square-law detection, we now introduce the deflection signal-to-noise ratio. If one were observing the output of a nonlinear detector on an oscilloscope, the appearance of the trace with the arrival of a signal at time t_0 would typically be as illustrated in Figure 6.8. Although, prior to t_0 the detector input consists of mean-zero Gaussian noise, because of the nonlinear operation the mean value at the output is no longer zero, which has been shown. Thus, prior to the time t_0, the trace a_n for noise only has some nonzero average value $\overline{a_n}$, and rms fluctuation

$$\sigma_a = (\overline{a_n^2} - \bar{a}_n^2)^{1/2} \tag{6.33}$$

and with the arrival of the signal the trace would be deflected to a value

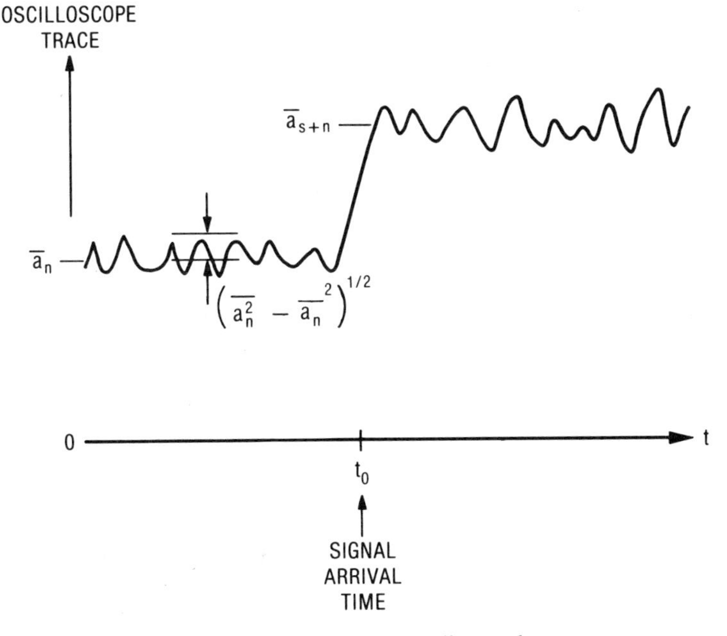

Figure 6.8 Output of non-linear detector.

a_{s+n}. For an observer who is looking for the deflection of the trace as an indication of the presence of a target (or a digital signal pulse) the quantity

$$\frac{\overline{a_{s+n}} - \overline{a_n}}{\sigma_a}$$

is a reasonable measure of the extent to which the target's presence would be detected. For example, a value of 10 would mean that the average deflection from $\overline{a_n}$ due to the presence of a target is ten-times the rms fluctuations of the trace due to noise.

In the case of interest here the observed quantity is the result of integration of M successive outputs of the square-law detector, given by $Z = \Sigma_{i=1}^{M} z^2(i)$ and the deflection signal-to-noise ratio DSNR [23] is defined here as

$$\text{DSNR} = \frac{\bar{Z}_{s+n} - \bar{Z}_n}{\sigma_Z} \tag{6.34}$$

where $\sigma_Z = (\overline{Z_n^2} - \bar{Z}_n^2)^{1/2}$. The quantity Z_{s+n} involves summation of square-law-detector outputs $z_{s+n}^2(i)$ when signal and noise are present, and Z_n denotes noise only. We now calculate DSNR. The ith observation $y_i(t)$ is

$$y_i(t) = AP_T(t)\cos(2\pi f_0 t + \theta_i) + n_i(t)$$

with the usual assumptions for $n(t)$, and using (6.10)

$$z_{s+n}^2(i) = \left[\int_0^T [A\cos(2\pi f_0\tau + \theta_i) + n_i(\tau)]\cos 2\pi f_0\tau\, d\tau\right]^2 + \left[\int_0^T [A\cos(2\pi f_0\tau + \theta_i) + n_i(\tau)]\sin 2\pi f_0\tau\, d\tau\right]^2 = I_i^2 + Q_i^2 \tag{6.35}$$

where in the interest of maximum generality the unknown signal phase θ_i is allowed to vary randomly and independently over $(0, 2\pi)$ between successive observations. The terms I_i and Q_i are explicitly

$$I_i = A\cos\theta_i \int_0^T \cos 2\pi^2 f_0\tau\, d\tau - A\sin\theta_i \int_0^T \sin 2\pi f_0\tau \cos 2\pi f_0\tau\, d\tau + \int_0^T n_i(\tau)\cos 2\pi f_0\tau\, d\tau$$

$$Q_i = A\cos\theta_i \int_0^T \cos 2\pi f_0\tau \sin 2\pi f_0\tau\, d\tau - A\sin\theta_i \int_0^T \sin 2\pi f^2 f_0\tau\, d\tau + \int_0^T n_i(\tau)\sin 2\pi f_0\tau\, d\tau \tag{6.36}$$

For the same reasons discussed in connection with (6.4) the integrals with integrands $\sin 2\pi f_0 \tau \cos 2\pi f_0 \tau$ are negligible, and also

$$\int_0^T \sin^2 2\pi f_0 \tau \, d\tau = \int_0^T \cos^2 2\pi f_0 \tau \, d\tau \approx \frac{T}{2}$$

Hence, as before, using the notation

$$X_i = \int_0^T n_i(\tau) \cos 2\pi f_0 \tau \, d\tau$$

$$Y_i = \int_0^T n_i(\tau) \sin 2\pi f_0 \tau \, d\tau \tag{6.37}$$

the ith output of the square-law detector when the signal is present is

$$\begin{aligned} z_{s+n}^2(i) &= \left(\frac{AT}{2} \cos \theta_i + X_i\right)^2 + \left(\frac{AT}{2} \sin \theta_i + Y_i\right)^2 \\ &= \frac{A^2T^2}{4} + AT(X_i \sin \theta_i + Y_i \cos \theta_i) + X_i^2 + Y_i^2 \end{aligned} \tag{6.38}$$

X_i and Y_i have the following properties. Referring to (6.14) et seq.

$$E(X_i^2) = E(Y_i^2) = \frac{N_0 T}{4} \tag{6.39}$$

Also by using the identity for Gaussian random variables ξ_i

$$\begin{aligned} E(\xi_1 \xi_2 \xi_3 \xi_4) &= E(\xi_1 \xi_2)E(\xi_3 \xi_4) + E(\xi_1 \xi_3)E(\xi_2 \xi_4) \\ &\quad + E(\xi_1 \xi_4)E(\xi_2 \xi_3) \end{aligned} \tag{6.40}$$

it is easily shown that (see Exercise 6.6).

$$E(X_i^4) = E(Y_i^4) = 3 \frac{N_0^2 T^2}{16} \tag{6.41}$$

and

$$E(X_i^2 Y_i^2) = E(X_i^2)E(Y_i^2) = \frac{N_0^2 T^2}{16} \tag{6.42}$$

That is, X_i^2 and Y_i^2 are uncorrelated. And by setting $A = 0$ in (6.38)

$$E[z_n^2(i)] = E(X_i^2 + Y_i^2) = \frac{N_0 T}{2} \tag{6.43}$$

With the use of these results, the numerator of (6.34) is

$$\bar{Z}_{s+n} - \bar{Z}_n = E\left[\sum_{i=1}^{M} z_{s+n}^2(i) - \sum_{i=1}^{M} z_n^2(i)\right] = \frac{MA^2T^2}{4} \tag{6.44}$$

The square of the denominator σ_Z^2 is

$$\sigma_Z^2 = E(Z_n^2) - [E(Z_n)]^2$$

$$= E\left[\left(\sum_{i=1}^{M} X_i^2 + Y_i^2\right)^2\right] - \left[E\sum_{i=1}^{M} (X_i^2 + Y_i^2)\right]^2 \tag{6.45}$$

$$= E\sum_i (X_i^2 + Y_i^2)\sum_j (X_j^2 + Y_j^2) - M^2\left(\frac{N_0T}{2}\right)^2 \tag{6.46}$$

where (6.39) has been used, and the summation terms reduce to

$$\sum_i (X_i^2 + Y_i^2)\sum_j (X_j^2 + Y_j^2)$$

$$= \sum_i X_i^4 + \sum_i\sum_j X_i^2X_j^2 + \sum_i Y_i^4 \quad (i \neq j)$$

$$+ \sum_i\sum_j Y_i^2Y_i^2 + 2\sum_i X_i^2 \sum_j Y_i^2 \quad (i \neq j) \tag{6.47}$$

In the double-summations in which $i \neq j$ there are $M(M-1)$ terms, in which X_i^2 and X_j^2 are independent, as are Y_i^2 and Y_j^2. Also by (6.42) X_i^2 and Y_i^2 are independent for all i. Equation (6.47) then becomes, using (6.39)–(6.44):

$$E\sum_i (X_i^2 + Y_i^2)\sum_j (X_i^2 + Y_i^2) = \frac{MN_0^2T^2}{4} + \frac{M^2N_0^2T^2}{4} \tag{6.48}$$

and using (6.46)

$$\sigma_Z^2 = \frac{MN_0^2T^2}{4} \tag{6.49}$$

from which, using (6.44), (6.34) is

$$\text{DSNR} = \frac{MA^2T^2/4}{(MN_0^2T^2/4)^{1/2}} = \sqrt{M}\,\frac{A^2T}{2N_0} = \sqrt{M}\,\frac{E}{N_0} \tag{6.50}$$

where $E = A^2T/2$ in the signal energy. Since in this scheme the integration takes place at the output of the detector, noncoherent integration is also referred to as post-detection integration, which yields an improvement in DSNR by a factor equal to the square root of the number of summations. For a linear detector it can be shown that the improvement also goes as $\sqrt{M}$. This case however is more cumbersome and is not worked out here.

6.3.2 Coherent Integration

According to [23] the first published report proposing the use of coherent integration was by Emslie [24]. In essence, the processor maintains the phase relationship between the received pulses so that the signal amplitude adds coherently. Since the noise is independent from pulse to pulse an improvement in effective signal-to-noise ratio over noncoherent integration can be achieved. The outputs of the integrator are usually input to, say, a square-law detector, and coherent integration is therefore also referred to as predetection integration. As has been noted, it is not necessary for the exact value of the signal phase to be known for this purpose. If this value is known some additional benefit can be obtained, exactly analogous to that obtained by coherent over noncoherent detection of a single pulse discussed in Section 6.2. In fact, it will be shown that this alternative is equivalent to implementing a filter that is matched to the entire pulse train, and in what follows we make the distinction between predetection integration, or coherent integration, and what might be termed generalized matched filtering. It will also be shown however that, practically speaking, there would be no significant advantage in doing this if increase in receiver sensitivity were the only purpose.

A receiver for implementing coherent integration is illustrated in Fig. 6.9. The input is a train of M pulses, each one of the form $P_T(t)\cos(2\pi f_0 t + \theta)$, where θ is assumed to be unknown. This might represent backscatter from a target or a repeated "1" in a binary communication channel. The oscillator used in generating the transmitted signal as well as the local oscillators in the receiver I and Q channels would ordinarily be derived from the same clock. Coherent operation requires that the clock maintain phase stability over time periods of the order of the duration of the M-pulse waveform or the round-trip travel time up to and back from the target, whichever is shorter.†

Noise is not shown in the figure, nor are the sum-frequency terms at the outputs of the multipliers, which would be eliminated by the IF stage. Also, the echo would in general have undergone a Doppler shift due to target motion that would have to be accommodated. This could be accomplished by employing a bank of such receivers, in which contiguous local oscillators would be offset in frequency by an amount equal to the width of a Doppler-resolution cell. In this way the entire range of possible Doppler shifts could be covered, and the range rate of the target determined by noting which of the receivers produces the maximum output. Alternatively, Doppler shifts could be accommodated by employing a large-BT Doppler-invariant waveform such as is described in Chapters 8 and 9. For purposes of simplicity in this presentation however, target motion is not considered here.

† Practically speaking, it is actually only necessary that the two local oscillator signals remain orthogonal over this time period.

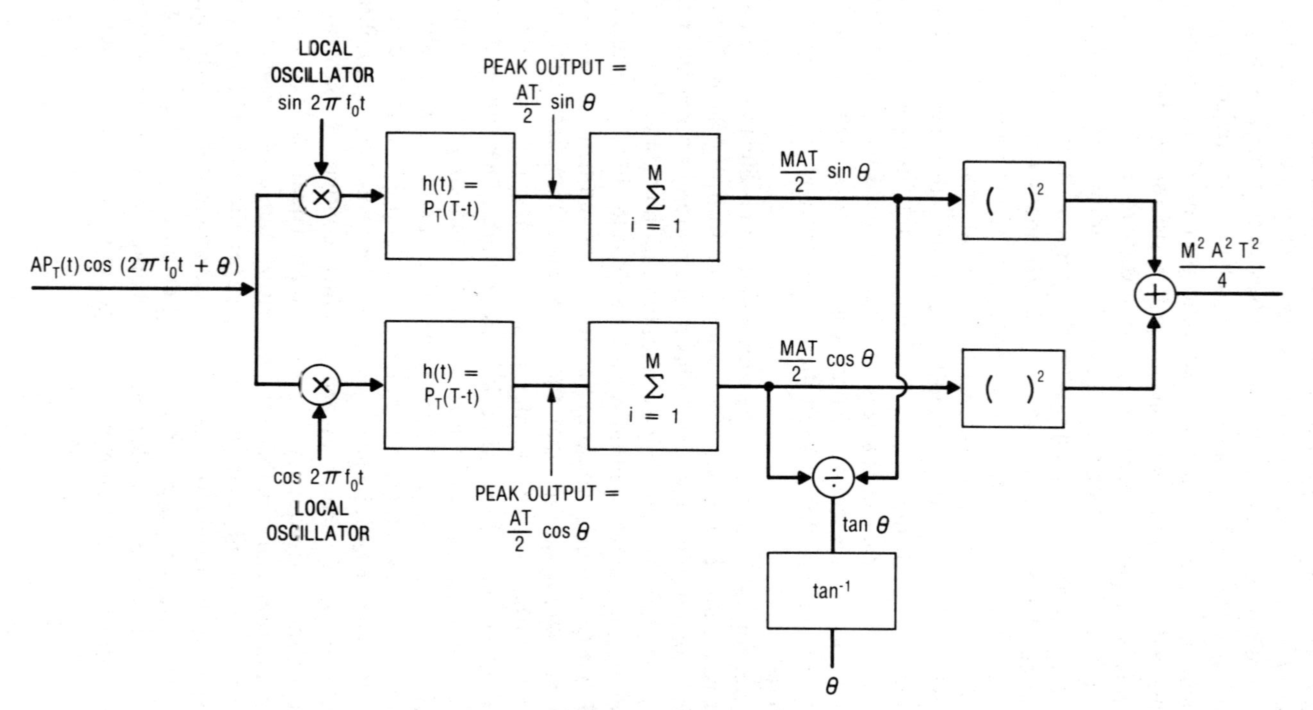

Figure 6.9 Receiver for coherent integration.

The outputs of the multipliers are passed through baseband filters matched to the video pulse shape, and the peak outputs of the filters in the in-phase and quadrature channels are $AT \cos \theta/2$ and $AT \sin \theta/2$ respectively. The integration which follows, by which SNR improvement is achieved, is accomplished by periodically overlapping and adding successive filter responses, the period being equal to the time separation between successive pulse transmissions. At this point the phase can be determined, if so desired, by dividing the quadrature term by the in-phase term and applying the arc-tangent operation. However, it is only necessary to add the outputs of the square-law detectors to acquire the coherently summed signal amplitude; measurement of θ is not necessary for this purpose. This is possible because $\sin^2\theta + \cos^2\theta = 1$ and a quadrature receiver must therefore be used. Eliminating the integrator in Figure 6.9—or equivalently setting $M = 1$—yields a receiver for measuring the phase of a single pulse, on the basis of which a coherent matched filter such as is discussed in Section 6.2 could be implemented.

In Figure 6.9 it has been tacitly assumed that the unknown phase θ remains constant over the duration of the M-pulse waveform, which (aside from deterministic variation due to target motion which we do not consider here) is generally assumed to be the case when coherent integration is employed. This assumption was not necessary in the preceding section for noncoherent integration. Here, however, it is clear that if this is not the case the signal will not add coherently from pulse to pulse and the process will not yield the desired SNR improvement.

Now, referring to Figure 6.9 and (6.13), at the instant of correlation for the ith signal pulse the outputs I_i and Q_i of the filters in the in-phase and quadrature channels are

$$\begin{aligned} I_i &= \frac{AT \cos \theta}{2} + \int_0^T n_i(\tau) \cos 2\pi f_0 \tau \, d\tau \\ &= \frac{AT \cos \theta}{2} + X_i \\ Q_i &= \frac{AT \sin \theta}{2} + \int_0^T n_i(\tau) \sin 2\pi f_0 \tau \, d\tau \\ &= \frac{AT \sin \theta}{2} + Y_i \end{aligned} \tag{6.51}$$

which becomes, after integration

$$\begin{aligned} \sum_{i=1}^{M} I_i &= \frac{MAT \cos \theta}{2} + \sum_{i=1}^{M} X_i \\ \sum_{i=1}^{M} Q_i &= \frac{MAT \sin \theta}{2} + \sum_{i=1}^{M} Y_i \end{aligned} \tag{6.52}$$

And after adding the outputs of the square-law detectors for the in-phase and quadrature channels

$$Z_{s+n} = \left(\frac{MAT\cos\theta}{2} + \sum_{i=1}^{M} X_i\right)^2 + \left(\frac{MAT\sin\theta}{2} + \sum_{i=1}^{M} Y_i\right)^2 \tag{6.53}$$

In order to calculate DSNR it is necessary to evaluate $E(Z_{s+n}^2)$, which is equal to

$$E(Z_{s+n}^2) = \frac{M^2A^2T^2}{4} + E\left[\left(\sum_i X_i\right)^2 + \left(\sum_i Y_i\right)^2\right]$$

because I_i and Q_i have zero mean, and

$$\overline{Z_{s+n}^2} - \overline{Z_n^2} = \frac{M^2A^2T^2}{4} \tag{6.54}$$

For the denominator of (6.34) we must calculate the square root of the quantity

$$E\left[\left(\sum_i X_i\right)^2 + \left(\sum_i Y_i\right)^2\right]^2 - \left[E\left(\sum_i X_i\right)^2 + E\left(\sum_i Y_i\right)^2\right]^2 \tag{6.55}$$

The first term is

$$E\left[\left(\sum_i X_i\right)^4 + \left(\sum_i Y_i\right)^4 + 2\left(\sum_i X_i\right)^2\left(\sum_i Y_i\right)^2\right] \tag{6.56}$$

By writing

$$\left(\sum_i X_i\right)^4 = \sum_i X_i \sum_j X_j \sum_k X_k \sum_l X_l \tag{6.57}$$

and using (6.40) it is easily shown that

$$\left(\sum_{i=1}^{M} X_i\right)^4 + \left(\sum_{i=1}^{M} Y_i\right)^4 = \frac{6M^2N_0^2T^2}{16} \tag{6.58}$$

The cross term in (6.56) is, using (6.42),

$$2E\left(\sum_i X_i\right)^2\left(\sum_i Y_i\right)^2 = 2E\left(\sum_i X_i^2 \sum_i Y_i^2\right) = \frac{2M^2N_0^2T^2}{16} \tag{6.59}$$

so that (6.56) is equal to $M^2N_0^2T^2/2$. Hence (6.55) is

$$\frac{M^2N_0^2T^2}{2} - \frac{M^2N_0^2T^2}{4} = \frac{M^2N_0^2T^2}{4} \tag{6.60}$$

and finally, from (6.34)

$$\text{DSNR} = \frac{M^2A^2T/4}{(M^2N_0^2T^2/4)^{1/2}} = \frac{MA^2T}{2N_0} = M\frac{E}{N_0} \tag{6.61}$$

Thus we have the important result that the improvement in SNR yielded by coherent integration is equal to the number of coherent additions M, whereas as seen in (6.50) for noncoherent or postdetection integration the improvement goes as $\sqrt{M}$, the square-root of the number of terms in the integration. But M is also the increase over a single pulse of the received signal energy. Thus the essential feature of coherent processing is exhibited; namely, that all the energy in the received signal is effectively recovered.

To achieve this it was only necessary to be differentially coherent from pulse to pulse. Now suppose θ were known, and consider a filter matched to each pulse, with impulse response: $h(t) = P_T(T-t)\cos(2\pi f_0(T-t)+\theta)$. It is not difficult (see Exercise 6.7) to show that in this case, in comparison with (6.51), the ith output of the in-phase channel at $t = T$ is $\frac{1}{2}AT + X_i$. Thus the signal is confined entirely within the in-phase channel, the quadrature channel need not be considered, and the integrated output is

$$\frac{MAT}{2} + \sum_{i=1}^{M} X_i \tag{6.62}$$

and using the definition of SNR of (5.61)

$$\text{SNR} = \frac{M^2A^2T^2/4}{E[\Sigma_{i=1}^{M} X_i]^2} = \frac{M^2A^2T^2}{MN_0T} = \frac{MA^2T}{N_0} = \frac{M2E}{N_0} \tag{6.63}$$

In fact, we have effectively implemented a generalized coherent matched filter to a waveform consisting of M repeated pulses, as evidenced by the factor of 2. The reason the integrated SNR in this case is twice that yielded by the quadrature detector, (6.61), is because the noise in the quadrature channel as well as the in-phase channel comes into play there. The situation is therefore an exact parallel to that discussed in Section 6.2, and it will be seen that the comparison in performance of coherent integration and generalized matched filtering also exactly parallels coherent and noncoherent detection of a single pulse.

6.4 PERFORMANCE OF COHERENT AND NONCOHERENT INTEGRATION

Let us now compare the performance of the integration schemes that have been considered, in terms of false alarm and detection probabilities.

6.4.1 Noncoherent Integration

Referring to (6.38), the output of the square-law detector can be written as

$$S_M = \sum_{i=1}^{M} \left(\frac{AT}{2}\cos\theta_i + X_i\right)^2 + \left(\frac{AT}{2}\sin\theta_i + Y_i\right)^2 \tag{6.64}$$

where X_i and Y_i are mean-zero Gaussian random variables with variance $N_0T/4$. The random variable S_M has a noncentral chi-square distribution with $2M$ degrees of freedom. For an exact solution the reader is referred to [21] which presents a concise and thorough discussion of chi-square together with exact performance calculations giving P_d vs P_f as a function of E/N_0 and M based on expansion of integrals of the chi-square distribution in a Gram–Charlier series. Here, a simplified approximate solution shall be presented which approaches the exact solution as M becomes large and in fact is accurate to 1 dB for $M \geq 4$ for typical situations of interest.

Equation (6.64) can be written as

$$S_M = \sum_{i=1}^{M} R_i \tag{6.65}$$

with

$$R_i = \left(\frac{AT}{2}\cos\theta_i + X_i\right)^2 + \left(\frac{AT}{2}\sin\theta_i + Y_i\right)^2$$

where the R_i are independent, identically distributed random variables. By the central-limit theorem therefore (Section 2.9), the probability density function of S_M approaches a Gaussian in the limit as $M \to \infty$ with mean μ_M and variance σ_M^2 given by

$$\mu_M = ME(R_i) = \frac{MA^2T^2}{4} + \frac{MN_0T}{2} = \frac{MN_0T}{2}\left(1 + \frac{E}{N_0}\right) \tag{6.66}$$

$$\sigma_M^2 = M[E(R_i^2) - (E(R_i))^2]$$

where the value of μ_M follows directly from (6.38). Calculation of σ_M^2 requires evaluating

$$\begin{aligned} E(R_i^2) &= E\left(\frac{A^2T^2}{4} + ATX_i\cos\theta_i + ATY_i\sin\theta_i + X_i^2 + Y_i^2\right)^2 \\ &= E(a+b+c+d+e)^2 \\ &= E(a^2+b^2+c^2+d^2+e^2+2ab+2ac+2ad+2ae \\ &\quad + 2bc+2bd+2be+2cd+2ce+2de) \end{aligned} \tag{6.67}$$

Equation (6.67) is simple to evaluate because the expected value of all cross terms involving odd powers of either X_i or Y_i vanish, and by using (6.39) and (6.41) (see Exercise 6.8),

$$\sigma_M^2 = \frac{MN_0^2T^2}{4}\left(1 + 2\,\frac{E}{N_0}\right) \tag{6.68}$$

Now for the false alarm probability set E equal to zero in (6.66) and (6.68), let $\mu = MN_0T/2$ and $\sigma^2 = MN_0^2T^2/4$, and the threshold η is determined from

$$P_{\mathrm{fa}} = \int_\eta^\infty e^{-(x-\mu)^2/2\sigma^2}\,\frac{dx}{(2\pi\sigma^2)^{1/2}} = \frac{1}{2}\,(1 - \operatorname{erf}(\gamma)) \tag{6.69}$$

where

$$\gamma = \left(\frac{\mu - \eta}{\sqrt{2}\sigma}\right)$$

and

$$\eta = \frac{M}{2}\,N_0T\left(1 + \gamma\sqrt{\frac{2}{M}}\right) \tag{6.70}$$

The detection probability is

$$\begin{aligned} P_{\mathrm{d}} &= \int_\eta^\infty e^{-(x-\mu_M)^2/2\sigma_M^2}\,\frac{dx}{\sqrt{2\pi}\sigma_M} \\ &= \frac{1}{2}\left\{1 + \operatorname{erf}\left[\frac{\left(\sqrt{\frac{M}{2}}\,\frac{E}{N_0} - \gamma\right)}{\left(1 + \frac{2E}{N_0}\right)^{1/2}}\right]\right\} \end{aligned} \tag{6.71}$$

which shows more directly the benefit in the improvement of SNR, and that P_{d} can be made arbitrarily close to unity, for any γ, if M can be made large enough.

A physical picture of the effect of M can be gained by noting that the separation of the peaks of the Gaussian distributions under H_0 and under H_1, is, from (6.66), equal to $ETM/2$. Also, the standard deviation under H_0 is $\sqrt{M}N_0T/2$. In order to achieve a low value of P_{fa} and a sufficiently large P_{d} it is necessary that the probability distributions under H_0 and H_1 be distinctly separated with as little overlap as possible. To achieve this it is necessary that $(ETM/2)\sqrt{M}N_0T/2 = \sqrt{M}E/N_0$ be large.

In order to determine the range of M for which (6.71) is a useful approximation, plots of P_{d} vs E/N_0 are presented in Figure 6.10 for different

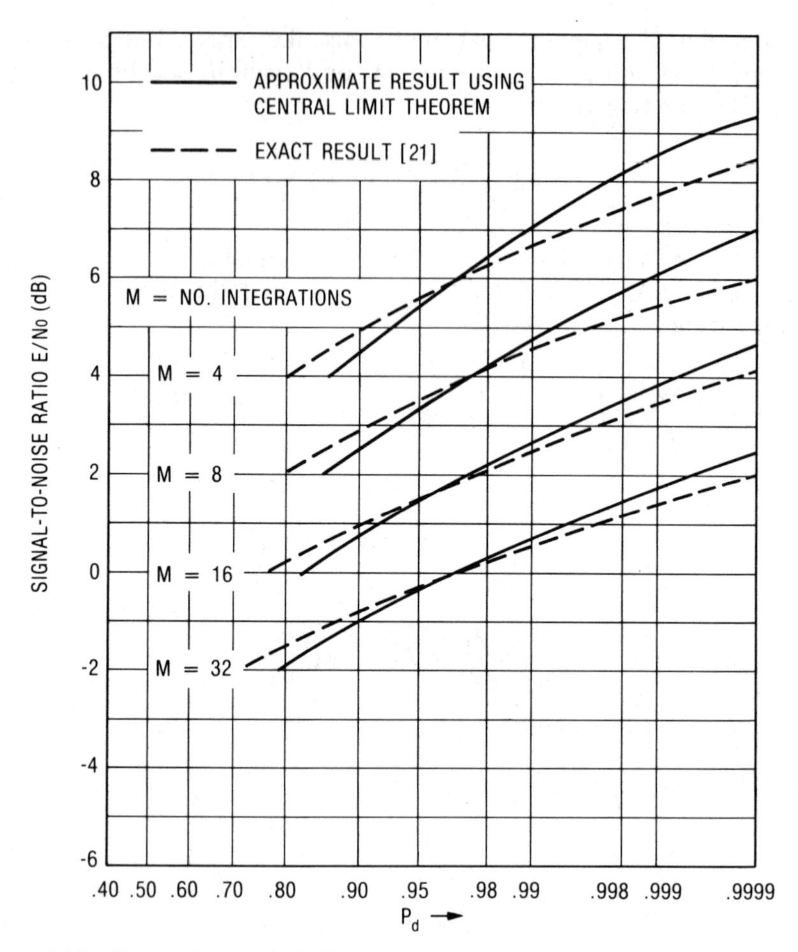

Figure 6.10 Comparison of SNR requirements with noncoherent integration ($P_{fa} = 10^{-2}$).

values of $M \geq 4$ using (6.71) along with more exact results from [21] which are based on the $2M$-degree-of-freedom noncentral chi-square distribution. The comparison in Figure 6.10 is for $P_{fa} = 10^{-2}$. In all cases the value of E/N_0 required to obtain a given value of P_d as determined using (6.71) is within 1 dB of the value obtained using the more exact solution.

6.4.2 Coherent Integration

In this case an exact solution is obtained. For predetection integration the output of the square-law detector is (6.53).

$$\left(\frac{MAT}{2}\cos\theta + \sum X_i\right)^2 + \left(\frac{MAT\sin\theta}{2} + \sum_{i=1}^{M} Y_i\right)^2 \qquad (6.72)$$

which has a noncentral chi-square distribution with two degrees of freedom. Now in general, a noncentral chi-square random variable z with N degrees of freedom is of the form

$$z = \sum_{i=1}^{N} (B_i + x_i)^2 \tag{6.73}$$

where the B_i are constants, the x_i are independent, mean-zero Gaussian, with $\mathrm{Var}(x_i) = \sigma^2$, and z has a probability density [21]

$$P(z) = \frac{1}{2\sigma^2}\left(\frac{z}{\beta}\right)^{(N-2)/4} \exp\left[-\frac{(z+\beta)}{2\sigma^2}\right] I_{N/2-1}\left[\frac{(z\beta)^{1/2}}{\sigma^2}\right] \tag{6.74}$$

where $\beta = \Sigma_{i=1}^{N} B_i^2$, and $I_{N/2-1}$ is the modified Bessel function of order $N/2-1$. In the case of interest here this takes a very simple form because $N=2$. Thus, $B_1 = \frac{1}{2}MAT\cos\theta$, $B_2 = \frac{1}{2}MAT\sin\theta$, $x_1 = \Sigma_{i=1}^{M} X_i$, $x_2 = \Sigma_{i=1}^{M} Y_i$, and x_1 and x_2 are each independent, mean-zero Gaussian with variance $\sigma^2 = MN_0T/4$. It can be shown using $I_0(0) = 1$ (see Exercise 6.9) that

$$P_{\mathrm{fa}} = \int_{\eta^2}^{\infty} \exp\left(-\frac{z}{2\sigma^2}\right)\frac{dz}{2\sigma^2} = \exp\left(-\frac{\eta^2}{2\sigma^2}\right) \tag{6.75}$$

where for the square-law detector the threshold has been defined as η^2, and

$$\begin{aligned} P_{\mathrm{d}} &= \int_{\eta^2}^{\infty} \exp\left(-\frac{z+\beta}{2\sigma^2}\right) I_0\left[\frac{(z\beta)^{1/2}}{\sigma^2}\right]\frac{dz}{2\sigma^2} \\ &= \int_{(-2\ln P_{\mathrm{fa}})^{1/2}}^{\infty} x\exp\left(-\frac{x^2+\alpha^2}{2}\right) I_0(x\alpha)\,dx \end{aligned} \tag{6.76}$$

where $\alpha^2 = 2ME/N_0$. Equations (6.75) and (6.76), in which no approximations have been made, are seen to be identical with (6.19) and (6.21) with $2E/N_0$ in the latter case being replaced by $2ME/N_0$. Thus there is an effective improvement in SNR by a factor of M which goes over directly into the calculation of P_{d}, and the curves of Figure 6.6 apply exactly with ME/N_0 substituted for E/N_0 on the vertical axis. This is the expected result. Coherent predetection integration increases SNR by a factor of M and, aside from this change, the subsequent nonlinear operation reproduces exactly the results of Section 6.2.

For generalized matched filtering, by (6.22), (6.23), (6.63) and (5.52)

$$P_{\mathrm{d}} = \frac{1}{2}\left[1 + \mathrm{erf}\left(\sqrt{\frac{2ME}{N_0}} - \gamma\right)\right] \tag{6.77}$$

where, as before, γ is determined from P_{fa} using

$$P_{\mathrm{fa}} = \frac{1}{2}[1 - \mathrm{erf}(\gamma)], \quad \eta = 2\gamma\sqrt{\frac{ME}{N_0}} \tag{6.78}$$

The latter two results, (6.75), (6.76) and (6.77), (6.78) illustrate that the difference between coherent-integration employing square-law detection and generalized matched filtering is identical to that between coherent and noncoherent detection of a single pulse as discussed in Section 6.2.

6.5 SUMMARY OF COHERENT AND NONCOHERENT DETECTION AND PROCESSING

Coherent detection of a single pulse requires employment of a coherent matched filter, for which the exact value of the phase of the input signal must be known. For noncoherent detection of a single pulse the filter bandwidth is matched to that of the signal but the phase is unknown. The difference in the two cases in terms of the transmitted pulse energy required to achieve a given value of P_d for a specified value of P_{fa} becomes significant (a ~3 dB advantage in the coherent case) only for small values of E/N_0—say in the vicinity of 3 dB. This occurs because in the coherent case the quadrature noise term is eliminated, whereas for noncoherent detection both the in-phase and quadrature noise terms come into play. However, for $E/N_0 > 10$ dB this becomes less important because the quadrature-channel noise under this condition has very little effect on the fluctuations of the output signal-plus-noise envelope (or envelope squared), and is therefore effectively eliminated. For such values of E/N_0 the difference in SNR required to achieve a given value of P_d for a fixed P_{fa} between coherent and noncoherent detection is about 1 dB, and eventually becomes negligible as E/N_0 increases further.

For a fixed value of single-pulse energy E the detection capability of a sensor can be increased by employing a waveform consisting of a train of M pulses and adding, or integrating, their cumulative responses. If integration is done after detection—postdetection or noncoherent integration—which destroys the phase relationship between successive pulses, the signal pulses add incoherently and the integrated SNR is increased by a factor of $\sqrt{M}$. However, if the phase relationship between pulses is maintained by employing stable local oscillators, and the integration is implemented prior to detection, the SNR is improved by a factor of M, which of course can amount to a very large difference (see Exercises 6.14, 6.15, 6.16). In this it is not necessary that the absolute value of the signal phase be known. The phase however could be measured and, ideally, with the use of this information a filter matched to the entire M-pulse waveform could be implemented. The relative advantage thereby obtained however is identical to that between coherent and noncoherent detection of a single pulse, of the order of 3 dB in transmitted signal energy for $ME/N_0 \leq 2$, and essentially negligible for $ME/N_0 > 10$. As a result, exact phase knowledge is not important in coherent integration, because integration is generally employed expressly for the purpose of achieving large values of ME/N_0. It should be noted however that there might be reasons other than signal detection for

TABLE 6.1

Single Pulse

COHERENT DETECTION

$$P_{\mathrm{fa}} = \frac{1}{2}[1 - \mathrm{erf}(\gamma)]$$

$$P_{\mathrm{d}} = \frac{1}{2}\left[1 + \mathrm{erf}\left(\sqrt{\frac{E}{N_0}} - \gamma\right)\right]$$

NONCOHERENT DETECTION

$$P_{\mathrm{d}} = \int_{(-2\ln P_{\mathrm{fa}})^{1/2}}^{\infty} \exp\left(-\frac{x^2+\alpha^2}{2}\right) I_0(x\alpha)\, dx$$
$$= Q(\alpha, -2\ln P_{\mathrm{fa}})$$

where $\alpha = 2E/N_0$ and $Q(\alpha, \beta)$ is Marcum's Q function

M-Pulse Waveform

NONCOHERENT INTEGRATION: APPROXIMATE SOLUTION FOR $M \geq 4$

$$P_{\mathrm{fa}} = \frac{1}{2}[1 - \mathrm{erf}(\gamma)]$$

$$P_{\mathrm{d}} = \frac{1}{2}\left\{1 + \mathrm{erf}\left[\frac{\sqrt{\frac{M}{2}}\frac{E}{N_0} - \gamma}{\left(1 + \frac{2E}{N_0}\right)^{1/2}}\right]\right\}$$

COHERENT INTEGRATION

$$P_{\mathrm{d}} = \int_{(-2\ln P_{\mathrm{fa}})^{1/2}}^{\infty} \exp\left(-\frac{x^2+\alpha^2}{2}\right) I_0(\alpha, x)\, dx$$
$$= Q(\alpha_M, -2\ln P_{\mathrm{fa}})$$

where $\alpha_M = ME/N_0$

Generalized Matched Filtering for M Pulses

$$P_{\mathrm{fa}} = \tfrac{1}{2}[1 - \mathrm{erf}(\gamma)]$$

$$P_{\mathrm{d}} = \frac{1}{2}\left[1 + \mathrm{erf}\left(\sqrt{\frac{ME}{N_0}} - \gamma\right)\right]$$

which knowledge of the signal phase might be of interest. This is discussed in Chapter 7.

A summary of the quantitative performance results of this chapter are presented in Table 6.1. Although a specific form of the signal was used, the results are perfectly general. This is demonstrated in Chapter 10, where the same results are obtained using a generalized complex signal. We note that calculations of P_d vs P_{fa} vs E/N_0 and M can in all cases be carried out by using Figure 6.6 and the table of values of the error function in the Appendix. For predetection integration, the vertical axis of Figure 6.6 should be interpreted as ME/N_0 rather than E/N_0.

EXERCISES FOR CHAPTER 6

6.1 A real bandpass function $x(t)$ can be written as $R[h(t)e^{i2\pi f_0 t}]$ where $h(t)$ can be complex. If $y(t) = \text{Re}[g(t)e^{i2\pi f_0 t}]$ show that

$$\int_{-\infty}^{\infty} x(t)y(t)\,dt = \frac{1}{2B}\sum_{n=-\infty}^{\infty}\left[h_R\left(\frac{n}{B}\right)g_R\left(\frac{n}{B}\right) + h_I\left(\frac{n}{B}\right)g_I\left(\frac{n}{B}\right)\right]$$

by making use of the carrier sampling theorem (4.15). It is assumed that $x(t)$ and $y(t)$ are observed at the output of a bandpass filter such as shown in Figure 6.1. Recall that the Nyquist rate here is $\Delta t = 1/B$.

6.2 For input $A \sin 2\pi f_0 t + B \cos 2\pi f_0 t$ determine the outputs of detectors in Figure 6.4a,b,c. Show how low-pass filtering or video amplification produces the desired results. Show that b produces a single harmonic at $2f_0$ whereas the others produce an infinite number of harmonics.

6.3 For $s(t) = P_T(t)\cos(2\pi f_0 t + \theta)$ show that the response of a matched filter appears as in Figure 6.5a, whereas for a filter matched in amplitude only the output appears as in Figure 6.5b (assume $\theta \neq 0$).

6.4 For input $A\cos(2\pi f_0 t + \theta)$ calculate the SNR as defined in (5.61) at the output of
a. a filter $h(t) = P_T(t)\cos[2\pi f_0(T - t) + \theta]$
b. a filter $h(t) = P_T(t)\cos 2\pi f_0(T - t)$
Assume noise to be white, Gaussian with PSD $N_0/2$.

6.5 The Rice distribution in (6.17) differs from that in (3.53) in that in (3.53) the variable r has the dimensions of volts whereas in (6.17) the dimension of z are volts $\times$ $T/2$ which is the effect of the filter. Find the appropriate transformation of the variable r so that (3.53) reduces to (6.17) and demonstrate this.

6.6 Show that $E(X_i^4) = E(Y_i^4) = 3N_0^2T^2/16$ where X_i and Y_i are given by (6.37). Using (6.40), the result of (6.4) and $E[n(t_1)n(t_2)] = \frac{1}{2}N_0\delta(t_2 - t_1)$, also show that X_i^2 and Y_i^2 are uncorrelated; that is $E(X_i^2Y_i^2) = E(X_i^2)E(Y_i^2)$.

6.7 Verify that the ith output of generalized matched filter is $\frac{1}{2}AT + X_i$; that is, the quadrature noise term vanishes.

6.8 Evaluate (6.67) and show that (6.66) reduces to (6.68).

6.9 Show that chi-square with two degrees of freedom reduces to the Rice distribution.

6.10 An observation takes place over 10 s. For $P_{fa} = 10^{-6}$ and PRF = 150 pulses/s write down an expression for the probability of k false alarms during this interval. What would be the expected value of k, and the mean time between false alarms. Repeat for $P_{fa} = 10^{-4}$ (see Section 11.4).

6.11 The dimensions of the threshold η for a matched filter and for ideal noncoherent detection are different. For $\frac{1}{2}N_0 = 10^{-6}$ W/Hz find the value of E for a coherent matched filter such that $P_{fa} = 10^{-4}$ can be obtained with a threshold of $\eta = 22$. What would be the resulting value of P_d?

For the same value of P_d and P_{fa} find the required value of E/N_0 for an envelope detector. Assume the noise is limited in bandwidth to the signal bandwidth. Find the required value of received signal power for a threshold of $\eta = 22$. Discuss the difference in units.

6.12 A target is to be observed for 10 s with a sensor operating with a PRF of 100. It is desired that the probability of more than one false alarm during this interval be $\leq 10^{-4}$. The detection probability is to be 0.998. Define the necessary system parameters (i.e. SNR and threshold) to achieve this for a coherent system (see Section 11.4).

6.13 For $E/N_0 = 20$ find P_d for $P_{fa} = 10^{-4}$ and 10^{-6} for single pulse coherent and noncoherent detection. Repeat for $E/N_0 = 30$. You will need tables of the Q function for this.

6.14 A system employs postdetection integration. What values of M are required to achieve a detection probability of $P_d = 0.99$ for $P_{fa} = 10^{-3}$ for values of E/N_0 of 0.25, 2, 10 and 100.

6.15 Repeat Exercise 6.13 assuming predetection integration and compare results.

6.16 Repeat Exercise 6.14 for generalized matched filtering and compare results.

7

PARAMETER ESTIMATION AND APPLICATIONS

This discussion is facilitated with the use of the analytic signal formulation. For this purpose we recall that generation of the detection statistic

$$\int_0^T y(t)s(t)\,dt$$

for a signal in Gaussian noise in (5.42), arises from the quadratic exponent in the numerator of (5.38) which, in the continuous-time representation, is, ignoring multiplicative constants,

$$\int_0^T [y(t) - s(t)]^2\,dt$$

Now, using (4.39)

$$\int_0^T (y(t) - s(t))^2\,dt = \frac{1}{2}\int_0^T |z_y(t) - z_s(t)|^2\,dt \tag{7.1}$$

where $z_y(t)$ and $z_s(t)$ are the analytic signals of $y(t)$ and $s(t)$. It therefore follows that

$$\int_0^T y(t)s(t)\,dt = \frac{1}{2}\,\mathrm{Re}\int_0^T z_y(t)z_s^*(t)\,dt \tag{7.2}$$

Recall that (7.2) can be generated by operating on the input with a matched filter with impulse response $h(t) = s(T - t)$, for which the output at time t for an input $y(t)$ commencing at $t = 0$ is

$$\int_0^t y(\tau)h(t-\tau)\,d\tau = \int_0^t y(\tau)s(T-t+\tau)\,d\tau \tag{7.3}$$

which is equal to (7.2) at $t = T$. Thus the impulse response of a matched filter in the analytic signal formulation is $z_s^*(T-t)$, and the complex output at time t for input $y(t)$ is

$$\frac{1}{2}\,z_y(t)^*z_h(t) = \frac{1}{2}\int_0^t z_y(\tau)z_s^*(T-t+\tau)\,d\tau \tag{7.4}$$

for which the real part at $t = T$ is equal to (7.2).

7.1 ESTIMATION OF RANGE TO A TARGET

The range to a target is to be measured using a transmitted signal

$$s(t) = \text{Re}[\psi(t)e^{i2\pi f_0 t}] = \psi_R(t)\cos 2\pi f_0 t - \psi_I(t)\sin 2\pi f_0(t) \tag{7.5}$$

with

$$\Psi(f) = \int_{-\infty}^{\infty} \psi(t)e^{-i2\pi f t}\,dt \tag{7.6}$$

The signal $s(t)$ is a generalized carrier pulse of duration T (Figure 7.1a), and $\psi(t)$ is a slowly varying complex signal. Although $s(t)$ and therefore $\psi(t)$ are time limited, the assumption is again made that all the signal energy outside some range $|f| > B/2$ is negligible, i.e. $\Psi(f) = 0$ for $|f| > B/2$; since we are interested in carrier signals with bandwidth B, their lowpass envelopes extend over $-B/2 \le f \le B/2$. Under this condition, for $f_0 > B/2$ the analytic signal $z_s(t)$ of $s(t)$, which is defined as $z_s(t) = s(t) + i\hat{s}(t)$, is under these circumstances also given by

$$z_s(t) = \psi(t)e^{i2\pi f_0 t} \tag{7.7}$$

That is, as is discussed in Section 4.4 (see Exercise 7.1), if $\psi(t)$ is bandlimited as defined above then the Hilbert Transform (HT) of $\psi(t)\cos 2\pi f_0 t$ is: $\text{HT}(\psi(t)\cos t2\pi f_0 t) = \psi(t)\,\text{HT}(\cos 2\pi f_0 t) = \psi(t)\sin 2\pi f_0 t$, and of course the same relationship holds for $\psi(t)\sin 2\pi f_0 t$, from which (7.7) follows.

The carrier frequency f_0 will be defined more explicitly below. The requirement $f_0 > B/2$ will be satisfied in all cases of practical interest since, as discussed in Chapter 3, unless this condition holds there will be an insufficient number of cycles within the pulse duration for us to speak of an amplitude-modulated carrier at all.

Now referring to (7.4), the analytic signal of the impulse response $h(t)$ of the filter matched to $s(t)$ is

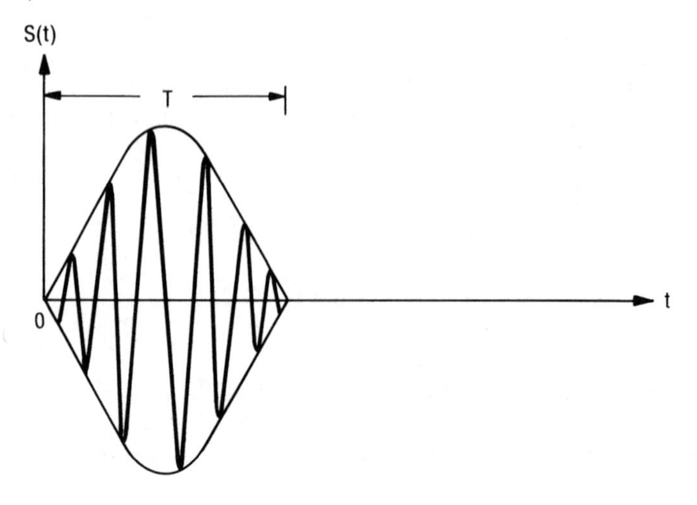

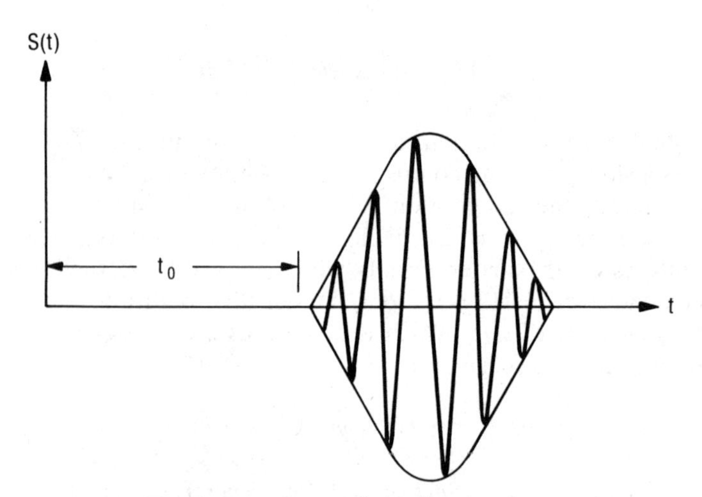

Figure 7.1 Generalized signal pulse.

$$z_h(t) = z_s^*(T - t) = \psi^*(T - t)e^{-i2\pi f_0(T-t)} \tag{7.8}$$

with

$$h(t) = \text{Re}[\psi^*(T - t)e^{-i2\pi f_0(T-t)}] \tag{7.9}$$

Also, the bandpass representation of Gaussian noise can be written as

$$n(t) = \text{Re}[\nu(t)e^{i2\pi f_0 t}] = n_c(t)\cos 2\pi f_0 t + n_s(t)\sin 2\pi f_0 t \tag{7.10}$$

where

$$\nu(t) = n_c(t) - in_s(t) \tag{7.11}$$
$$z_n(t) = \nu(t)e^{i2\pi f_0 t}$$

where $z_n(t)$ is the analytic signal of the noise, which is also band-limited.

The signal $s(t)$ is scattered by the target and the echo arrives back at the sensor at some time t_0 after transmission, as illustrated in Figure 7.1b. We wish to detect the presence of the echo and measure, or more precisely estimate, the target's range $r = ct_0/2$ where c is the speed of signal propagation. The time t_0 is determined by noting the time that $s(t)$ is transmitted—which in what follows shall for convenience be defined as $t = 0$—and the time that the peak output of the receiver is observed and making an appropriate subtraction. The received echo will of course have been attenuated in propagation. This however can be ignored in what follows if the signal (7.6) is redefined explicitly such that

$$\int_{-\infty}^{\infty} |\psi(t)|^2\, dt = \int_{-\infty}^{\infty} |\Psi(f)|^2\, df = 2E \tag{7.12}$$

where E is the received rather than the transmitted signal energy. Any amplitude or phase distortion in the signal due to target motion,[†] dispersion, multipath, etc., is also ignored. For the noise, if the signal bandwidth is B it will be assumed that the in-band noise power accompanying the signal is $N_0 B$ where $N_0 = kT$ (3.83).

Chapter 6, which deals with coherent and noncoherent operations, discusses various receiver configurations. Specifically, Section 6.3.2 discusses how phase information in a signal can be preserved, which is of interest for Section 7.3 which shows how phase information can be used in the measurement of range. It is however more customary in practice to estimate range from observation of the peak response of a square-law or envelope detector in a receiver such as that diagrammed in Figure 6.3, which employs an incoherent matched filter. In what follows both coherent and noncoherent matched filtering shall be considered, with square-law detection in the latter case which greatly simplifies the analysis and is negligibly different from envelope detection for these purposes.

These alternatives are represented symbolically in Figure 7.2 by the observation points A and B. Observations at A represent coherent operation, and observation at B represents noncoherent operation. Either case admits the possibility of pre- or postdetection integration to improve SNR.

[†] To be discussed in Chapter 8. These effects are ignored here.

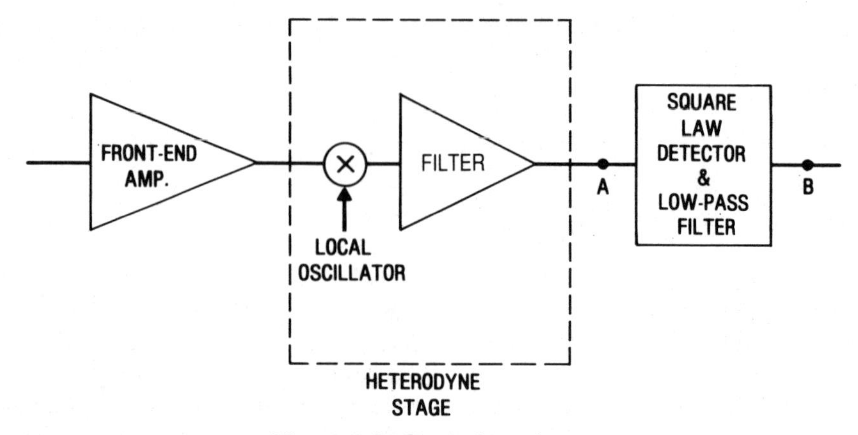

Figure 7.2 Typical receiver.

Referring to (7.4), (7.7) and (7.8) the complex output at point A if the filter were coherently matched to the signal is

$$\begin{aligned}\frac{1}{2}\int_{-\infty}^{\infty} z_s(\tau)z_h(t-\tau)\,d\tau &= \frac{1}{2}\int_{-\infty}^{\infty} \psi(\tau-t_0)e^{i2\pi f_0(\tau-t_0)}\psi^*(T-t+\tau) \\ &\quad\times e^{-i2\pi f_0(T-t+\tau)}\,d\tau \\ &= \frac{1}{2}\,e^{i2\pi f_0(t-T-t_0)}\int_{-\infty}^{\infty}\psi(\tau-t_0)\psi^*(T-t+\tau)\,d\tau \end{aligned} \tag{7.13}$$

and the output of the square-law detector at point B is

$$f(t) = \frac{1}{4}\left|\int_{-\infty}^{\infty}\psi(\tau-t_0)\psi^*(T-t+\tau)\,d\tau\right|^2 \tag{7.14}$$

where infinite limits can be used since the definition of $\psi(t)$ includes its finite duration T. In (7.13) and in what follows, for simplicity of notation we continue to use f_0 to denote the signal carrier frequency, and omit making the distinction between the transmitted carrier frequency and the intermediate frequency (IF), which would be different from f_0 in a heterodyne receiver.

Notice in (7.13) and (7.14) that the additional terms that normally arise in the real-signal formulation, which we generally ignore by making use of the approximation of (6.4), do not arise here. *This follows from the foregoing band-limited assumption; the approximation of* (6.4) *is equivalent to assuming the signal to be band-limited as discussed above* (*see Exercise* 7.4).

In (7.13), if the filter were matched in amplitude only the phase in the exponential multiplying the integral in (7.13) would contain an arbitrary unknown phase θ. But this of course makes no difference in the detector outputs at point B and (7.14) gives the output of the square-law detector for either case. Also, the detector output will be zero until $t \geq t_0$, at which time $f(t)$ begins to increase, and the maximum is reached at $t_M = t_0 + T$, which is easily proved by applying the Schwarz inequality (5.66). Again, this is a result of the band-limited approximation, since the additional terms that are encountered with real signals, however small, which are eliminated by the approximation of (6.4), would result in the detector output not being necessarily maximum at $t = t_0 + T$ if the filter were not coherently matched (see Exercise 7.5).

In any case, the observable of interest is the time at which the maximum detector output occurs, from which, under the band-limited approximation, and for simplicity ignoring delays in waveguides, cables, etc. t_0 is determined from

$$t_0 = t_M - T \tag{7.15}$$

In practice the approximation $t_0 \approx t_M$ is often made. However, if very accurate measurements are required the delay T imposed by the filter should be included, as well as all other delays in the processing path. At $t = t_M$, (7.14) is

$$f(t_M) = \frac{1}{4} \left| \int_{-\infty}^{\infty} |\psi(t)|^2 \, dt \right| = \frac{1}{4} \left| \int_{-\infty}^{\infty} |\Psi(f)|^2 \, df \right|^2 = E^2 \tag{7.16}$$

where E is the energy in the received signal and, for simplicity, the multiplicative effects of amplifier gain, losses in waveguides and cables, etc. have been ignored. Since these shall also be ignored for the noise, which is affected by them equally, the ratio E/N_0 will remain unchanged.

Now let us consider the effect of noise on the measurement. As will be seen shortly, if target range is to be measured to satisfactory accuracy a reasonably large signal-to-noise ratio is required. In this case, although detection of the echo will not be difficult, some perturbation of the pulse shape by the noise can be expected, with the result that the location of the peak of the detector output may be displaced from its position in the noise-free case; also, since the filter bandwidth is matched to the signal bandwidth the noise at the filter output cannot fluctuate more rapidly than the signal. To a first approximation therefore, as is illustrated in Figure 7.3, let us assume that the output of the square-law detector in the noise free case $f(t)$, becomes transformed in the presence of noise to $f(t - \epsilon)$, in which case $\sigma_\epsilon = (\overline{\epsilon^2})^{1/2}$ is a measure of the accuracy with which the range r to the target can be measured.

In calculating σ_ϵ it is convenient to use the following explicit definition of carrier frequency f_0 [25], in terms of the spectrum $Z_s(f)$ of the analytic signal $z_s(t)$

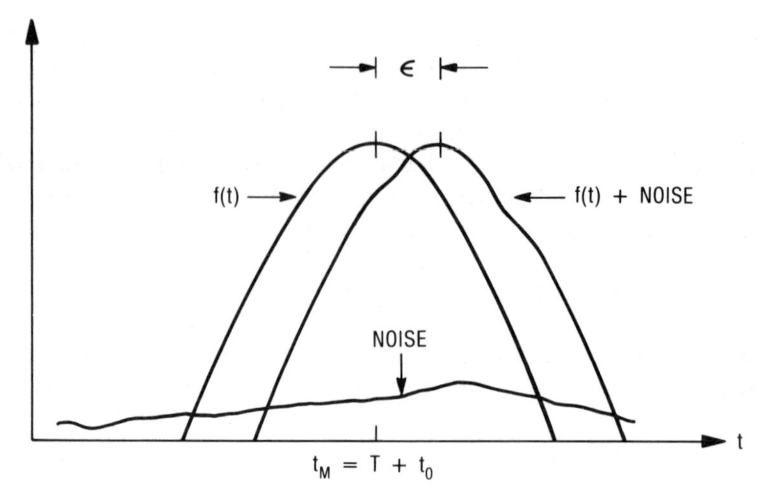

Figure 7.3 Effect of noise on range measurement.

$$Z_s(f) = \int_{-\infty}^{\infty} z_s(t) e^{-i2\pi ft}\, dt \tag{7.17}$$

which has only positive frequencies. The carrier frequency is defined as the distance of $Z_s(f)$ from the origin, in terms of the centroid of $|Z_s(f)|^2$, as

$$f_0 = \frac{\int_{-\infty}^{\infty} f|Z_s(f)|^2\, df}{\int_{-\infty}^{\infty} |Z_s(f)|^2\, df} = \frac{\int_{-\infty}^{\infty} f|Z_s(f)|^2\, df}{2E} \tag{7.18}$$

But from (7.7)

$$Z_s(f) = \Psi(f - f_0) \tag{7.19}$$

where $\Psi(f)$ is given by (7.6). Hence, by substituting (7.19) and (7.16) into (7.18),

$$\int_{\infty}^{\infty} f|\Psi(f)|^2\, df = 0 \tag{7.20}$$

Thus this definition of f_0 defines the low-pass envelope functions of $s(t)$, $\psi_R(t)$ and $\psi_I(t)$, such that the centroids of their spectra around $f = 0$ vanish. If, as is often the case, $|s(t)|$ is symmetrical about its midpoint, this means that $|\Psi(f)| = |\Psi(-f)|$, and $|Z_s(f)|$ will therefore also be symmetrical around $f = f_0$.

We also introduce the quantity β [25] defined by

$$\begin{aligned}\beta^2 &= \frac{\int_{-\infty}^{\infty} [2\pi(f - f_0)]^2 |Z_s(f)|^2\, df}{2E} \\ &= \frac{\int_{-\infty}^{\infty} (2\pi f)^2 |\Psi(f)|^2\, df}{2E}\end{aligned} \tag{7.21}$$

which is essentially the center of gravity of $|Z_s(f)|^2$ about the carrier frequency f_0, or equivalently, of the low-pass function $|\Psi(f)|^2$ about $f = 0$. As such, β is obviously related to the bandwidth of $s(t)$, but the specific percentage of the total signal energy in $|\Psi(f)|^2$ contained within the range $-\beta/2 \leq f \leq \beta/2$ will depend on the particular pulse shape $s(t)$. For a signal with a rectangular spectrum B it is found (see Exercise 7.6) that $\beta = \pi B/\sqrt{3} \approx 1.8B$.

From these definitions of f_0 and β the following relationships are easily shown to hold (see Exercise 7.7).

$$\int_{-\infty}^{\infty} \psi^*(t) \frac{d\psi(t)}{dt}\, dt = \int_{-\infty}^{\infty} \psi^*(t)\psi'(t)\, dt = 0 \tag{7.22}$$

$$\beta^2 = \frac{\int_{-\infty}^{\infty} \psi^*(t)\psi''(t)\, dt}{\int_{-\infty}^{\infty} |\psi(t)|^2\, dt} \tag{7.23}$$

where prime denotes derivative with respect to the variable of integration, t. Now denote

$$\zeta(t) = \int_{-\infty}^{\infty} \psi(\tau - t_0)\psi^*(T - t + \tau)\, d\tau \tag{7.24}$$

and in the noise-free case the output of the square-law detector $f(t)$ is $\frac{1}{4}|\zeta(t)|^2$, and at $t = t_M$

$$f(t_M) = \frac{1}{4}|\zeta(t_M)|^2 = \frac{1}{4}\left|\int \psi(\tau)\psi^*(\tau)\, d\tau\right|^2 = E^2 \tag{7.25}$$

Referring to (7.10) the input to the filter when noise is present is $z_y(t) = z_s(t - t_0) + z_n(t)$ which can also be written as $z_s(t - t_0) + z_n(t - t_0)$, which simplifies the calculation and sacrifices no generality because $z_n(t)$ is stationary and we shall ultimately be taking expected values. Thus in the presence of noise the output of the filter is

$$\frac{1}{2} e^{i2\pi f_0(t-T-t_0)}\left[\zeta(t) + \int_{-\infty}^{\infty} \nu(\tau - t_0)\psi^*(T - t + \tau)\, d\tau\right] \tag{7.26}$$

The output of the square-law detector is therefore

$$f(t) = \frac{1}{4}|\zeta(t)|^2 + \frac{1}{2}\operatorname{Re}\zeta^*(t)\int_{-\infty}^{\infty} \nu(\tau - t_0)\psi^*(T - t + \tau)\, d\tau + \frac{1}{4}\left|\int_{-\infty}^{\infty} \nu(\tau - t_0)\psi^*(T - t + \tau)\, d\tau\right|^2 \tag{7.27}$$

Hence by the foregoing discussion concerning the effect of noise on $f(t)$ the output of the square-law detector of time $t = t_M$ can be written as

$$f(t_M - \epsilon) = f(t_M) - \epsilon f'(t_M) + \tfrac{1}{2}\epsilon^2 f''(t_M) + \cdots$$
$$= f(t_M) + \frac{1}{2}\,\mathrm{Re}\,\zeta^*(t_M)\int_{-\infty}^{\infty} \nu(\tau)\psi^*(\tau)\,d\tau + \frac{1}{4}\left|\int_{-\infty}^{\infty} \nu(\tau)\psi^*(\tau)\,d\tau\right|^2 \tag{7.28}$$

where, again, the argument of $\nu(t)$ has been simplified by stationarity.

We want $\sigma_\epsilon^2 = E(\epsilon^2)$, and in taking the expected value of both sides of (7.28), $E(\nu(\tau)) = 0$, and also $E(\epsilon) = 0$ since ϵ can clearly be positive or negative (also $f'(t_M) = 0$). Hence ignoring terms higher than $\mathcal{O}(\epsilon^2)$:

$$\sigma_\epsilon^2 = \frac{\int_{-\infty}^{\infty}\int_{-\infty}^{\infty} E[\nu(\tau_1)\nu^*(\tau_2)]\psi(\tau_1)\psi^*(\tau_2)\,d\tau_1\,d\tau_2}{2f''(t_M)} \tag{7.29}$$

By using (7.22) and (7.23) it can be shown (see Exercise 7.8) that

$$2f''(t_M) = \frac{1}{2}\left.\frac{d^2|\zeta(t)|^2}{dt^2}\right|_{t=t_M} = 4\beta^2E^2 \tag{7.30}$$

For the numerator of (7.29), by (7.10) and (7.11),

$$E[\nu(\tau_1)\nu^*(\tau_2)] = E[n_c(\tau_1)n_c(\tau_2) + n_s(\tau_1)n_s(\tau_2)]$$
$$+ iE[n_c(\tau_1)n_s(\tau_2) - n_s(\tau_1)n_c(\tau_2)] = 2R(\tau_1 - \tau_2) \tag{7.31}$$

where $R(\tau_1 - \tau_2) = E[n_c(\tau_2)n_c(\tau_1)] = E[n_s(\tau_2)n_s(\tau_1)]$.

Now the signal contends with the noise waveform (7.10) whose power spectral density is equal to $N_0/2$ over two bands of width, say, B_N centered at $\pm f_0$. The correlation function of $n(t)$ is therefore

$$\frac{N_0}{2}\int_{-f_0-B_N/2}^{-f_0+B_N/2} e^{i2\pi f\tau}\,df + \frac{N_0}{2}\int_{f_0-B_N/2}^{f_0+B_N/2} e^{i2\pi f\tau}\,df = N_0\,\frac{\sin \pi B_N\tau}{\pi\tau}\,\cos 2\pi f_0\tau \tag{7.32}$$

But from (7.10) the correlation function of $n(t)$ is $\mathrm{R}(\tau)\cos 2\pi f_0\tau$. The correlation function of the slowly-varying functions $n_c(t)$ and $n_s(t)$ is therefore $N_0(\sin \pi B_N\tau/\pi\tau)$ which yields $R(0) = E(n^2(t)) = E(n_c^2(t)) = E(n_s^2(t)) = N_0B$ as expected. Thus the input bandpass waveform (7.10) has energy spread over two bands $|f \pm f_0| \leq B_N$ with power spectral density $N_0/2$, and the low-pass functions $n_c(t)$ and $n_s(t)$ have power spectral density N_0 over $|f| \leq \pm B_N/2$. The numerator of (7.29) is therefore, if $B_N \geq B$,

$$2N_0\int_{-B_N/2}^{B_N/2}\left|\int_{-\infty}^{\infty}\psi(\tau)e^{-i2\pi f\tau}\,d\tau\right|^2 df = 2N_0\int_{-B/2}^{B/2}|\Psi(f)|^2\,df = 4N_0E \tag{7.33}$$

Thus, for (7.33) to hold the bandwidth of the noise incident at the filter must at least equal the bandwidth of the filter. In physical terms, the foregoing result requires that successive noise samples separated in time by $1/B$ be uncorrelated, which will not be the case if $B_N < B$. Assuming this, substitution of (7.30) and (7.33) into (7.29) yields the result

$$\sigma_\epsilon \approx \frac{1}{\beta(E/N_0)^{1/2}} \tag{7.34}$$

Note that, as is discussed at the end of Sec. 6.2, if the filter were matched in phase as well as amplitude, the signal would contend with only the in-phase noise component. Hence, from (7.31), $R(\tau)$ rather than $2R(\tau)$ would be used in (7.33) and $2E/N_0$ would replace E/N_0 in (7.34).

The relationship between measurement accuracy, signal-to-noise ratio and resolution capability was first obtained in this form by Woodward [26]. Since $\beta = 1.8B$ for a rectangular bandwidth B, range-measurement accuracy increases with signal bandwidth and signal-to-noise ratio. Bandwidth is of course related to range-resolution capability. It is easily shown that two point scatterers separated in range by a distance $\Delta r + \delta$ will produce two distinct echoes, just time-resolved, if illuminated by a rectangular pulse of duration $T = 2\,\Delta r/c$; the separation between the echoes will be $2\delta/c$. The resolution capability of a rectangular pulse of duration T is therefore $\Delta r = cT/2$ or $1/B \sim T = 2\,\Delta r/c$. Using (7.34), the standard deviation of the range-measurement error $\sigma_r = c\sigma_\epsilon/2$ can be expressed in terms of the range-resolution width Δr as

$$\sigma_r = \frac{c/2}{1.8B\sqrt{E/N_0}} = \frac{cT/2}{1/8\sqrt{E/N_0}} = \frac{\Delta r}{1.8\sqrt{E/N_0}} \tag{7.35}$$

Equation (7.35) emphasizes the difference between the capability for distinguishing the presence of two separate closely spaced objects, and measuring the position of a single object when only one is present. This distinction has been confused by many, including, evidently, Lord Rayleigh [27]. While measurement accuracy σ_r is proportional to resolution capability Δr, it can in fact be thousands or even millions of times greater (better), depending on the value of E/N_0.

As is discussed above, this analysis applies to observations at the output of a square-law or envelope detector whether the filter prior to the detector is matched in phase and amplitude, or amplitude only. It also applies to outputs of a receiver employing predetection or postdetection integration. In this case however, as is discussed in Chapter 6, for pre-detection integration the factor E/N_0 in (7.34) or (7.35) would be replaced by ME/N_0, where M is the number of pulses integrated, and for post-detection integration E/N_0 would be replaced by $\sqrt{M}E/N_0$.

7.2 GENERALIZED PARAMETER ESTIMATION

In the previous section, based on physical considerations, an expression has been derived which gives the accuracy with which target range can be measured, and which identifies the system parameters on which the accuracy depends. It is of interest to ask whether this result represents a limit or whether one could possibly do better. In what follows the question of parameter estimation is addressed more generally, in terms of statistical estimation theory, in which it will be shown that (7.34) in fact essentially represents a bound on the accuracy that can be achieved, which follows from a general result with a wide range of applicability.

We first review briefly some elements of statistical estimation theory. The joint probability density function of N independent observations of a Gaussian random variable $y(t_i) = n(t_i) + s(t_i - t_0)$ with mean value $s(t_i - t_0)$, which represents the echo of a target at range $r = ct_0/2$ is

$$\prod_{i=1}^{N} \frac{\exp[-(y(t_i) - s(t_i - t_0))^2/2\sigma^2]}{(2\pi)^{N/2}\sigma^N}$$

In general, the joint, a priori conditional density function of N observations $y_1, y_2, \ldots, y_N$, from which an unknown parameter θ[†] is to be estimated, which in the above example would be t_0, is referred to as the likelihood function $L(y_1, y_2, \ldots, y_N; \theta)$. For N independent observations this takes the form

$$L(y_1, y_2, \ldots, y_N; \theta) = f(y_i, \theta) f(y_2; \theta) \cdots f(y_N; \theta) \tag{7.36}$$

In order to estimate θ an estimator $\hat{\theta}(y_1, y_2, \ldots, y_N)$ is constructed, which is a function of the N observations $y(t_1), y(t_2), \ldots, y(t_N)$, whose value after the observation is taken to be the estimate of the unknown parameter θ. Since the $y_i s$ are random variables the estimator is also a random variable. Thus any estimate of θ is a random quantity, and the mean and variance of $\hat{\theta}$ are a measure of how good an estimate of θ the estimator $\hat{\theta}$ actually is.

As an example, suppose the y_i, $i = 1, 2, \ldots, N$, are independent observations of a random variable whose unknown mean μ is to be estimated. A possible estimator of $\theta = \mu$ is

$$\hat{\theta}(y_1, y_2, \ldots, y_N) = \frac{1}{N} \sum_{i=1}^{N} y_i \tag{7.37}$$

[†] There should be no confusion between this standard usage of θ as an unknown parameter in estimation theory and previous usage in which θ represents phase of a sinusoid.

whose expected value is

$$E[\hat{\theta}] = \frac{1}{N} \sum_{i=1}^{N} E[y_i] = \mu \tag{7.38}$$

in which case the estimator is said to be *unbiased*, since its expected value is equal to the true value.

7.2.1 The Cramer–Rao Lower Bound on the Variance of an Estimator

It is clearly desirable for an estimator to be unbiased. For an unbiased estimator it is also desirable for the variance of the estimator $E(\hat{\theta} - \theta)^2$ to be as small as possible, since the fluctuations of any particular realization $\hat{\theta}(y_1, y_2, \ldots, y_N)$ about the true value $E(\hat{\theta})$ can then be expected to be correspondingly small. It is left as an exercise (Exercise 7.9) to show that if $\hat{\theta}$ is not unbiased, then $E(\hat{\theta} - \theta)^2$ is not the variance of the estimator $\hat{\theta}$. An important result will now be proved which establishes a lower bound on the variance of any estimator, thereby setting a limit on the accuracy that can be achieved.

Since $L(y_1, y_2, \ldots, y_N, \theta)$ is the joint density of $y_1, y_2, \ldots, y_N$, then from the foregoing discussion

$$\begin{aligned} E[\hat{\theta}(y_1, y_2, \ldots, y_N)] &= \int \hat{\theta}(y_1, y_2, \ldots, y_N) L(y_1, y_2, \ldots, y_N, \theta)\, dy_1\, dy_2 \cdots dy_N \\ &= \int \hat{\theta}(\mathbf{y}) L(\mathbf{y}, \theta)\, d\mathbf{y} = \theta + b(\theta) \end{aligned} \tag{7.39}$$

where the bias $b(\theta)$ can in the most general case be a function of the unknown parameter θ, and for convenience we write the N-tuple $(y_1, y_2, \ldots, y_N)$ as a vector $\mathbf{y}$.

It then follows that

$$\frac{\partial E[\hat{\theta}(\mathbf{y})]}{\partial \theta} = \int \hat{\theta}(\mathbf{y}) \frac{\partial L(\mathbf{y}, \theta)}{\partial \theta}\, d\mathbf{y} = 1 + \frac{\partial b(\theta)}{\partial \theta} \tag{7.40}$$

and also

$$\begin{aligned} &\int L(\mathbf{y}, \theta)\, d\mathbf{y} = 1 \\ &\int \frac{\partial L(\mathbf{y}, \theta)}{\partial \theta}\, d\mathbf{y} = 0 \\ &\int \theta \frac{\partial L(\mathbf{y}, \theta)}{\partial \theta}\, d\mathbf{y} = 0 \end{aligned} \tag{7.41}$$

Therefore, since $\partial L(\mathbf{y},\theta)/\partial\theta = L(\mathbf{y},\theta)[\partial \log L(\mathbf{y},\theta)/\partial\theta]$ it follows that

$$\int [\hat{\theta}(\mathbf{y}) - \theta] L(\mathbf{y},\theta) \frac{\partial \log L(\mathbf{y},\theta)}{\partial\theta}\, d\mathbf{y} = 1 + \frac{\partial b(\theta)}{\partial\theta} \tag{7.42}$$

But by the Schwarz inequality (5.66), since $L(\mathbf{y},\theta) = (L(\mathbf{y},\theta))^{1/2}(L(\mathbf{y},\theta))^{1/2}$

$$\left[\int [\hat{\theta}(\mathbf{y}) - \theta] L(\mathbf{y},\theta) \frac{\partial \log L(\mathbf{y},\theta)}{\partial\theta}\, \mathbf{dy}\right]^2$$

$$= \left[1 + \frac{\partial b(\theta)}{\partial\theta}\right]^2 \le \int [\hat{\theta}(\mathbf{y}) - \theta]^2 L(\mathbf{y},\theta)\, \mathbf{dy} \times \int L(\mathbf{y},\theta)\left[\frac{\partial \log L(\mathbf{y},\theta)}{\partial\theta}\right]^2 \mathbf{dy}$$

Hence

$$\Delta^2(\theta) \ge \frac{\left[1 + \frac{\partial b(\theta)}{\partial\theta}\right]^2}{\int L(\mathbf{y},\theta)\left[\frac{\partial \log L(\mathbf{y},\theta)}{\partial\theta}\right]^2 \mathbf{dy}} \tag{7.43}$$

which is the Cramer–Rao [28] lower bound on the quantity $\Delta^2(\theta) = E(\hat{\theta} - \theta)^2 = \int (\hat{\theta}(\mathbf{y}) - \theta)^2 L(\mathbf{y},\theta)\, \mathbf{dy}$ which is not the variance of the estimator $\hat{\theta}$ unless $b = 0$. If however $b = 0$ (7.43) yields the lower bound on the variance $\sigma^2_{\hat{\theta}}$ of an unbiased estimator $\hat{\theta}$

$$\sigma^2_{\hat{\theta}} \ge \frac{1}{\int L(\mathbf{y},\theta)\left[\frac{\partial \log L(\mathbf{y},\theta)}{\partial\theta}\right]^2 \mathbf{dy}} = \frac{1}{E\left[\left(\frac{\partial \log L(\mathbf{y},\theta)}{\partial\theta}\right)^2\right]} \tag{7.44}$$

Nowhere in this derivation has (7.36) been used, and (7.44) is therefore a general result in terms of the joint density function $L(\mathbf{y},\theta)$ of the $y_1, y_2, \ldots, y_N$ which need not be independent. If the y's are independent however (7.36) holds and the Cramer–Rao lower bound then becomes

$$\sigma^2_{\hat{\theta}} \ge \frac{1}{NE\left[\left(\frac{\partial \log f(y,\theta)}{\partial\theta}\right)^2\right]}$$

Equations (7.43) and (7.44) represent the limit on the performance of any estimator. An estimate that meets the Cramer–Rao lower bound is said to be *efficient*. If the variance of the (unbiased) estimator meets this lower bound as $N \to \infty$ the estimator is said to be *asymptotically efficient*. A third important property is sufficiency, which we do not deal with in any detail in what follows but discuss briefly here for completeness. An estimator is said to be *sufficient* if it makes use of all the information in the observations $(y_1, y_2, \ldots, y_n)$ concerning the value of the unknown parameter θ. A test for sufficiency is that the likelihood function can be written in the form

$$L(y_1, y_2, \ldots, y_n; \theta) = g(y_1, y_2, \ldots, y_n)h(\hat{\theta}, \theta) \tag{7.45}$$

To see why $\hat{\theta}$ is a sufficient statistic if (7.45) holds, note that in general $L(\mathbf{y}; \theta)$ can be written using Baye's rule in terms of the conditional density of $\mathbf{y}$ given the random variable $\hat{\theta}$ as

$$L(\mathbf{y}; \theta) = P_\theta(\mathbf{y}|\hat{\theta})P_\theta(\hat{\theta})$$

Now if the functional form of $\hat{\theta}$ is such that (7.45) holds, then by (2.17)

$$P_\theta(\mathbf{y}|\hat{\theta}) = \frac{L(\mathbf{y}; \theta)}{P_\theta(\hat{\theta})} = \frac{g(\mathbf{y})h(\theta, \hat{\theta})}{\int g(\mathbf{y})h(\theta, \hat{\theta})\, dy} = \frac{g(\mathbf{y})h(\theta, \hat{\theta})}{h(\theta, \hat{\theta})} = g(\mathbf{y})$$

Hence, if (7.45) holds, the conditional density of $\mathbf{y}$ given the values of $\hat{\theta}$ and θ is $g(\mathbf{y})$, which is independent of $\hat{\theta}$ and θ. Thus all the information concerning θ is contained in the function $h(\hat{\theta}, \theta)$ and no other estimator can yield any more information concerning θ; hence $\hat{\theta}$ is sufficient.

7.2.2 Maximum Likelihood Estimation

Having briefly summarized the properties of estimators we now deal with the question of how the estimator is to be constructed. The notion of maximum likelihood has already been encountered in Chapter 5. There it is used as a criterion for choosing between the two hypotheses H_0 and H_1. Here it is used for purposes of estimation. In this case the estimate $\hat{\theta}$ of θ is that value which maximizes the likelihood function $L(\mathbf{y}; \theta)$. That is, $\hat{\theta}$ satisfies

$$\left.\frac{\partial L(\mathbf{y}, \theta)}{\partial \theta}\right|_{\theta = \hat{\theta}} = \frac{\partial L(\mathbf{y}, \hat{\theta})}{\partial \theta} = 0 \tag{7.46}$$

Since the likelihood function $L(\mathbf{y}, \theta)$ is in fact also the conditional density of $\mathbf{y}$ given θ, the Maximum Likelihood Estimate (MLE) assigns to the unknown parameter θ that value $\hat{\theta}(\mathbf{y})$ which guarantees that this conditional density function will be maximized for any set of observations $\mathbf{y}$. This rationale for choosing an estimator is similar to that used for maximizing the probability of correct decision, as discussed in connection with (5.3).

As examples, consider a set of independent observations y_i, $i = 1, \ldots, N$ where y is Gaussian with known variance σ^2 and unknown mean μ which is the parameter to be estimated. The likelihood function is then

$$L(\mathbf{y}, \mu) = \frac{\exp\left[-\sum_{i=1}^{N} \frac{(y_i - \mu)^2}{2\sigma^2}\right]}{(2\pi)^{N/2}\sigma^N}$$

and (7.46) becomes

$$-L(\mathbf{y}, \mu) \frac{\partial}{\partial \mu} \sum_{i=1}^{N} \frac{(y_i - \mu)^2}{2\sigma^2} = L(\mathbf{y}, \mu) \sum_{i=1}^{N} \frac{(y_i - \mu)}{\sigma^2} = 0$$

which is satisfied for that value of the estimator $\hat{\mu}$ given by

$$\hat{\mu}(\mathbf{y}) = \frac{1}{N} \sum_{i=1}^{N} y_i \tag{7.47}$$

It can also be shown (see Exercise 7.10) that (1): if μ is known the MLE $\hat{\sigma}^2$ of σ^2 is

$$\hat{\sigma}^2 = \frac{1}{N} \sum_{i=1}^{N} (y_i - \mu)^2 \tag{7.48}$$

and if both μ and σ are unknown the unbiased MLE of σ^2 is

$$\hat{\sigma}^2 = \frac{1}{N-1} \sum_{i=1}^{N} (y_i - \hat{\mu})^2 \tag{7.49}$$

where $\hat{\mu}$ is given by (7.47). Also, (2): both (7.47) and (7.49) are unbiased. Both (7.47) and (7.49) are arithmetic averages. The reason for the factor $1/N-1$ in (7.49) is that for any set of observations $\mathbf{y}$ the estimate $\hat{\mu}$ is a constraint on the estimate of σ^2 for which there are therefore only $N-1$ independent quantities remaining. Therefore if the normalization is $1/N$ the estimate will be biased.

For purposes of estimation the method of maximum likelihood is very useful. In many cases the MLE satisfies the Cramer–Rao lower bound. Furthermore, it can be shown that the MLE will be an efficient and/or a sufficient estimate of an unknown parameter if in fact an efficient and/or a sufficient estimate for that parameter exists.

7.3 APPLICATIONS OF MAXIMUM LIKELIHOOD PARAMETER ESTIMATION TO SENSOR MEASUREMENTS

Referring to Section 5.1, the likelihood function in continuous-time representation for an observation of bandpass signals $y(t) = n(t) + s(t)$, with the usual assumptions for $n(t)$, is

$$\frac{\exp\left[-\frac{1}{N_0}\int_{-\infty}^{\infty}[y(t)-s(t)]^2\,dt\right]}{(2\pi)^{N/2}\sigma^N} = \frac{\exp\left[-\frac{1}{2N_0}\int_{-\infty}^{\infty}|z_y(t)-z_s(t)|^2\,dt\right]}{(2\pi)^{N/2}\sigma^N} \tag{7.50}$$

where $z_y(t)$ and $z_s(t)$ are the corresponding analytic signals. In the case of interest the transmitted signal is a pulsed carrier waveform such as (7.5), the received signal is $s(t-t_0)$ and t_0 is the unknown parameter representing target range which is to be estimated. Then the likelihood function $L(y(t), t_0)$ is

$$L(y(t), t_0) = \frac{1}{2\pi^{N/2}\sigma^N}\left\{\exp\left(-\frac{1}{2N_0}\int_{-\infty}^{\infty}|z_y(t)|^2\,dt\right)\exp\left(-\frac{E}{N_0}\right)\right.$$
$$\left.\exp\left[\frac{1}{N_0}\operatorname{Re}\int_{-\infty}^{\infty} z_y(t)z_s^*(t-t_0)\,dt\right]\right\} \tag{7.51}$$

where E is signal energy, and the MLE $\hat{t}_0$ of the unknown parameter t_0 is that value of t_0 which maximizes

$$\frac{\operatorname{Re}}{N_0}\int_{-\infty}^{\infty} z_y(t)z_s^*(t-t_0)\,dt \tag{7.52}$$

But, referring to (7.4) this is equivalent to processing $z_y(t)$ with a matched filter with complex impulse response $z_h(t) = z_s^*(T-t)$ for which the output at time t is

$$\int_{-\infty}^{\infty} z_y(\tau)z_s^*(T-t+\tau)\,d\tau \tag{7.53}$$

which yields (7.52) for $t = t_0 + T$. The value of t that maximizes (7.53) is therefore equivalent to the value of t_0 that maximizes (7.52). The estimate $\hat{t}_0$ which maximizes (7.52) is that value

$$\hat{t}_0 = t_M - T \tag{7.54}$$

which is obtained at the instant that the matched filter output is maximum at $t = t_M$. Thus, the matched filter yields the MLE of range.†

Moreover, the estimate is unbiased because (7.53) is

$$\int_{-\infty}^{\infty} z_n(\tau)z_s^*(T-t+\tau)\,d\tau + \int_{-\infty}^{\infty} z_s(\tau-t_0)z_s^*(T-t+\tau) \tag{7.55}$$

which for $z_n(t) = 0$ is maximum for $t = t_0 + T$. When noise is present (7.54) becomes

$$\hat{t}_0 = t_M - T + \epsilon \tag{7.56}$$

† This holds for the non-coherent as well as the coherent case—see (7.15).

where ϵ is a random perturbation. However on average the perturbation will clearly be symmetrical about the true value; hence $E(\epsilon) = 0$ and (7.44) applies.

7.3.1 Calculation of the Cramer–Rao Bound for Coherent and Noncoherent Observations

In calculating the Cramer–Rao lower bound there are two possibilities that will be considered. The range estimate can be based on coherent observations at the output of a matched filter, in which case the phase information in the signal can be utilized. Secondly, the estimate can be based on observations at the output of a square-law detector (or an envelope detector), which was the case dealt with in Section 7.1 and which shall be considered first.

Here the phase information is destroyed and all the available information resides in the position of the signal envelope, or envelope squared. In this case however, because of the non-linear operation the likelihood function would have to be formulated in terms of the chi-square or the Rice distribution and the calculation would be quite tedious. Let us avoid this by dealing with a hypothetical situation in which the transmitted signal is a video pulse of bandwidth B. This will yield essentially the same result, since the estimate uses no phase information, being based on information concerning the position of the envelope only, and permits a much simpler calculation because Gaussian statistics apply. Therefore assume the received signal to be a real, band-limited video pulse $v(t - t_0)$ with

$$v(t) = \int_{-B}^{B} V(f) e^{i2\pi ft}\, df$$

$$E = \int_{-\infty}^{\infty} v^2(t)\, dt = \int_{-B}^{B} |V(f)|^2\, df$$

$$\beta^2 = \frac{4\pi^2 \int_{-B}^{B} f^2 |V(f)|^2\, df}{E} \tag{7.57}$$

where, again, B is sufficiently large to include essentially all the signal energy and,

$$\int_{-\infty}^{\infty} v(t) v'(t)\, dt = 0 \tag{7.58}$$

for which symmetry of $v(t)$ about its midpoint is a sufficient condition. The likelihood function (7.50) is thus

$$L(\mathbf{y}, t_0) = \frac{\exp\left[-\dfrac{1}{N_0} \displaystyle\int_{-\infty}^{\infty} [y(t) - v(t - t_0)]^2\, dt\right]}{(2\pi)^{N/2} \sigma^N} \tag{7.59}$$

from which, using (7.58)

$$\frac{\partial \log L(\mathbf{y}, t_0)}{\partial t_0} = \frac{2}{N_0} \int_{-\infty}^{\infty} y(t) \frac{dv(t - t_0)}{dt_0} dt$$
$$= -\frac{2}{N_0} \int_{-\infty}^{\infty} y(t) \frac{d}{dt} v(t - t_0)\, dt = -\frac{2}{N_0} \int_{-\infty}^{\infty} y(t) v'(t - t_0)\, dt \tag{7.60}$$

and

$$E\left[\frac{\partial \log L(\mathbf{y}, t_0)}{\partial t_0}\right]^2 = \frac{4}{N_0^2} \int_{-\infty}^{\infty} \int_{-\infty}^{\infty} E[y(t_1)y(t_2)] v'(t_1 - t_0) v'(t_2 - t_0)\, dt_1\, dt_2 \tag{7.61}$$

Now

$$E[y(t_1)y(t_2)] = E[(n(t_1) + v(t_1 - t_0))(n(t_2) + v(t_2 - t_0))]$$
$$= E[n(t_1)n(t_2)] + v(t_1 - t_0)v(t_2 - t_0) \tag{7.62}$$

since $E[n(t_1)] = E[n(t_2)] = 0$, and when (7.62) is substituted into (7.61) the contribution from $v(t_1 - t_0)v(t_2 - t_0)$ vanishes by (7.58). Therefore using (7.31), (7.32)† and (7.57), (7.61) becomes

$$\frac{2}{N_0} \int_{-\infty}^{\infty} \int_{-\infty}^{\infty} \frac{\sin 2\pi B_N (t_1 - t_2)}{\pi (t_1 - t_2)} v'(t_1 - t_0) v'(t_2 - t_0)\, dt_1\, dt_2$$
$$= \frac{2}{N_0} \int_{-B_N}^{B_N} \left| \int v'(t - t_0) e^{2\pi f t}\, dt \right|^2 df$$
$$= \frac{2}{N_0} \int_B^B 4\pi^2 f^2 |V(f)|^2\, df = \frac{2\beta^2 E}{N_0} \tag{7.63}$$

if, as in (7.33), $B_N > B$, and therefore from (7.44)

$$\sigma_{\hat{t}_0} \geq \frac{1}{\beta (2E/N_0)^{1/2}} \tag{7.64}$$

Comparison of (7.64) with (7.34) show that the Cramer–Rao bound is achieved with a matched filter. Equation (7.64) has a wide range of applicability for estimation in the presence of Gaussian noise and applies to any measurement system in which the desired information resides in the location of the peak of some response. For example, range-rate can be determined from Doppler shift which can be estimated from the displace-

† The integration in this case however extends from $-B_N$ to B_N.

ment from the transmitted carrier frequency of the peak of the Fourier transform of the received echo; this is discussed in Chapter 8. Or the angular location of a target can be estimated by noting the azimuthal direction of a scanning search-radar antenna at which the peak response of a target occurs. In the former case, for a rectangular pulse of duration T, $\beta = 1.8T$. In the second case, for a linear radiator (e.g. antenna) of length L with a uniform excitation (e.g. current distribution, see Chapter 11) over L, $\beta = 1.8L/\lambda$ where λ is the wavelength of the transmitted signal. These values of β are calculated using (7.57) with the proper interpretation of $V(f)$ (see Exercise 7.11). In each of these examples $1/\beta$ is proportional to what might be termed the measurement-resolution width of the system. For Doppler measurements the frequency resolution width is $\sim 1/T$ and for an antenna, as is shown in Chapter 11, the angle-measurement resolution is λ/L. Hence, denoting Δ as the measurement-resolution width of any system, which is a measure of the capability of the system for resolving two closely spaced responses, the lower bound on the standard deviation $\sigma_{\hat{\theta}}$ of an unbiased estimator can be written as

$$\sigma_{\hat{\theta}} \geq \frac{k\Delta}{(2E/N_0)^{1/2}} \tag{7.65}$$

where k is a constant of order unity and $k\Delta$ can be calculated using (7.57) appropriately.

We turn now to the question of the bound on the variance of the estimator when phase information is utilized. From (7.50)–(7.52) and (7.60), returning to the use of analytic signals and, referring to the paragraph preceding (7.26), writing for convenience $z_n(\tau) = z_n(\tau - t_0)$,

$$\begin{aligned} -\frac{\partial \log L(y(t), t_0)}{\partial t_0} &= \frac{\text{Re}}{N_0} \int_{-\infty}^{\infty} z_y(\tau) z_s'(\tau - t_0)\, d\tau \\ &= \frac{\text{Re}}{N_0} \int_{-\infty}^{\infty} \nu(\tau - t_0) e^{i2\pi f_0(\tau - t_0)} \frac{d}{d\tau} \psi^*(\tau - t_0) e^{-i2\pi f_0(\tau - t_0)}\, d\tau \\ &\quad + \frac{\text{Re}}{N_0} \int_{-\infty}^{\infty} z_s(\tau - t_0) z_s'^*(\tau - t_0)\, d\tau \end{aligned} \tag{7.66}$$

The integrand of the last term on the right-hand side of (7.66) can be written as

$$\begin{aligned} \frac{1}{2} \frac{d}{d\tau} |z_s(\tau - t_0)|^2 &= \frac{1}{2} \frac{d}{d\tau} |\psi(\tau - t_0)|^2 \\ &= \frac{1}{2} \psi^*(\tau - t_0)\psi'(\tau - t_0) + \frac{1}{2} \psi(\tau - t_0)\psi'^*(\tau - t_0) \end{aligned} \tag{7.67}$$

Hence the last term on the right-hand side of (7.66) vanishes by (7.22). The first term is[†]

$$\frac{\text{Re}}{N_0}\int_{-\infty}^{\infty} \nu(\tau - t_0)[\psi'^*(\tau - t_0) - i2\pi f_0\psi^*(\tau - t_0)]\, d\tau$$

$$= \frac{\text{Re}}{N_0}[A + B] = \frac{1}{2N_0}[A + B + A^* + B^*] \tag{7.68}$$

We must calculate

$$\frac{1}{4N_0^2} E[A + B + A^* + B^*]^2$$

$$= \frac{E}{4N_0^2}[A^2 + B^2 + A^{*2} + B^{*2} + 2AB + 2A^*B^* + 2|A|^2 + 2|B|^2 + 2AB^* + 2BA^*] \tag{7.69}$$

But

$$E(A^2) = E(B^2) = E(A^{*2}) = E(B^*)^2) = 2E(AB) = 2E(A^*B^*) = 0$$

because $E(\nu(\tau_1)\nu(\tau_2)) = E(\nu^*(\tau_1)\nu^*(\tau_2)) = 0$ which is easily seen using (7.10) and (7.11). It is also easily shown that $AB^* = -BA^*$. The only nonvanishing terms are $2|A|^2$ and $2|B|^2$ and therefore

$$E\left[\frac{\partial \log L(y(t), t_0)}{\partial t_0}\right]^2$$

$$= \frac{1}{2N_0^2}\int_{-\infty}^{\infty}\int_{-\infty}^{\infty} E[\nu(\tau_1 - t_0)\nu^*(\tau_2 - t_0)]\psi'^*(\tau_1 - t_0)\psi'(\tau_2 - t_0)\, d\tau_1\, d\tau_2$$

$$+ \frac{1}{2N_0^2} 4\pi^2 f_0^2 \int_{-\infty}^{\infty}\int_{-\infty}^{\infty} E[\nu(\tau_1 - t_0)\nu^*(\tau_2 - t_0)]\psi^*(\tau_1 - t_0) \times \psi(\tau_2 - t_0)\, d\tau_1\, d\tau_2 \tag{7.70}$$

Hence by using

$$\psi'(\tau - t_0) = i2\pi \int_{-\infty}^{\infty} f\Psi(f)e^{i2\pi f(\tau - t_0)}\, df \tag{7.71}$$

and following exactly the same steps used in calculating the numerator of (7.29)

[†] $A = \int_{-\infty}^{\infty} \nu(\tau - t_0)\psi'^*(\tau - t_0)\, d\tau$.

$$E\left[\frac{\partial \log L(y(t), t_0)}{\partial t_0}\right]^2 = \frac{1}{N_0}\int_{-B_N/2}^{B_N/2} (4\pi^2 f^2 + 4\pi^2 f_0^2)|\Psi(f)|^2 \, df$$
$$= \frac{2E}{N_0}(\beta^2 + 4\pi^2 f_0^2) \tag{7.72}$$

if, as before, $B_N \geq B$. Therefore

$$\sigma_{\hat{t}_0} \geq \frac{1}{(\beta^2 + 4\pi^2 f_0^2)^{1/2}(2E/N_0)^{1/2}} \tag{7.73}$$

Referring to the discussion leading to (7.65) it is seen that the effective resolution of the system has increased from $\sim 1/\beta$ to $\sim 1/(\beta^2 + 4\pi^2 f_0^2)^{1/2}$ which can be a very large difference depending on the relative values of f_0 and β. It is easy to see physically why this should be the case. In a fully coherent system comparison of the phases of the transmitted and received signals can ideally resolve the distance to the target to a fraction of a wavelength. As a practical matter, since the target will be many wavelengths distant there is of course an ambiguity problem in identifying which particular wavelength is being subdivided. If however one wishes to measure only relative changes in position from pulse to pulse rather than the absolute value the ambiguity problem can be avoided.

Increase in system resolution with the use of phase information is illustrated in Figure 7.4 which shows the magnitude of a typical carrier

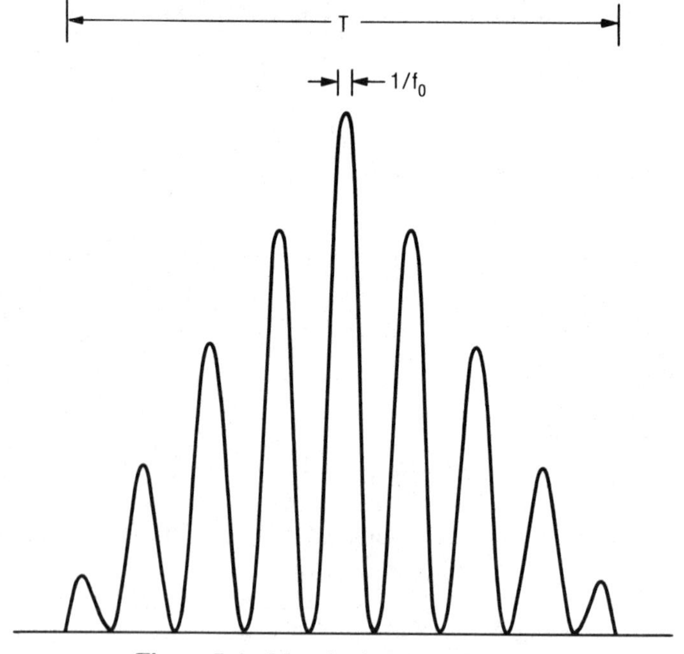

Figure 7.4 Magnitude of carrier pulse.

pulse, including the carrier. If the phase information is used, the system resolution is determined by the width of each carrier cycle, $\sim 1/f_0$, rather than by the width of the envelope $1/\beta$. For a rectangular bandwidth B, (7.73) can be rewritten as

$$\sigma_{\hat{t}_0} \gtrsim \frac{1}{3B(1+\pi^2 f_0^2/B^2)^{1/2}(2E/N_0)^{1/2}} = \frac{1}{1.8B(1+\pi^2 N^2)^{1/2}(2E/N_0)^{1/2}} \tag{7.74}$$

where (B/f_0) is the fractional bandwidth of the system which is nominally equal to the reciprocal of the number of carrier cycles per pulse, N. This shows that the possible advantages which could be gained in this way are potentially much greater for radar and laser radar than for sonar. In the former case, say for radar, typical parameters might be $f_0 = 1.5 \times 10^9$ Hz, $B = 10^8$ Hz, for which the fractional bandwidth is 0.06 or equivalently ~ 15 cycles per pulse; for laser radar N could be much larger. For sonar however, the fractional bandwidths are inherently larger because of the lower signal frequencies and there is less advantage to be gained; the same of course can hold true for modern radars if large fractional bandwidths are employed.

7.4 APPLICATION OF PARAMETER ESTIMATION TO TRACKING AND PREDICTION

Let us now apply the foregoing results to the tracking problem, which is of considerable practical importance in many applications of active sensing systems. Let the position of a target as a function of time be given by

$$z(t) = s + vt + \tfrac{1}{2}at^2 \tag{7.75}$$

where in general s, v and a are unknown. A sensor observes the target for a time duration $\sim(M-1)T_{\mathrm{p}}$ during which M pulses are transmitted, which yield M independent observations of the target positions equally spaced by the pulse-separation time T_{p}.

On the basis of these measurements we want estimates of s, v and a, an estimate of the target position at the end of the tracking time, and we want to be able to predict the position of the target at some time into the future after the track has ended. If there were no noise, ideally only three observations would be necessary to determine s, v, a and the position of the target would be known subsequently for all time. Because of noise however, the estimates of s, v and a will be random variables. Hence, surrounding the estimated target position at the end of the track, as well as the predicted position at some time in the future, there will be an error volume whose size represents the uncertainty in the estimated and predicted position due to noise. The question is then, how does the accuracy of the estimates of s, v,

a, target position, and predicted target position depend on the relevant system parameters. These parameters are the duration of the tracking time, or equivalently, the number of pulses (observations) transmitted during the tracking interval and the interpulse spacing, and E/N_0.

In typical tracking systems the problem of estimation of these quantities is usually dealt with by means of tracking algorithms employing Kalman filters or some variation thereof, which operate in real time on the measured data. In the development of such algorithms for a particular application however, it is essential to have some a priori understanding of the problem in terms of the required track duration and the number of pulses that will be required, which is gained by calculations of this kind.

The tracking scenario and the tracking waveform are illustrated in Figure 7.5. The time origin is referred to the center of the observation interval T and for convenience M is assumed to be odd. Thus the observation takes place over the interval

$$-\left(\frac{M-1}{2}\right)T_{\mathrm{p}} \leq t \leq \left(\frac{M-1}{2}\right)T_{\mathrm{p}} \tag{7.76}$$

The three geometrical coordinates chosen are range r and two orthogonal cross-range dimensions $x = r\theta_x$ and $y = r\theta_y$ where θ_x and θ_y are the cross-range angles. Measurements of r therefore correspond to measuring the time of arrival of received echoes as described earlier in this chapter and measurements of x and y correspond to the angular direction (θ_x, θ_y) in which the sensor is pointed as a function of time during the observation period. Equation (7.65) applies to each coordinate and there are actually nine quantities: (s_r, s_x, s_y), (v_r, v_x, v_y) and (a_r, a_x, a_y) to be estimated. Each coordinate however can be dealt with independently by separate application of (7.75), since they have been chosen to be orthogonal.

Now suppose we have M measurements of the form[†]

$$y(t_k) = z(t_k) + \eta(t_k)\,, \quad k = 1, 2, \ldots, M$$

where $z(t_k) = s + vt_k + \frac{1}{2}at_k^2$ can represent either r, x or y, and $\eta(t_k)$ is the error in the measurement due to noise which is assumed to have zero mean, and variance $E[\eta^2(t_k)] = \sigma^2$. Chang and Tabaczynski [29] have calculated unbiased estimates of s, v and a, together with the variances of the estimates, based on a weighted-least-square polynomial fit to noisy data. The results are found to be in very good agreement with Monte Carlo simulations employing typical tracking algorithms, and are therefore typical of what would be encountered in practice. Let us first consider for simplicity a case in which a is known exactly so that only s and v are to be estimated.

[†] $y(t_k)$ should not be confused with the cross-range coordinate y.

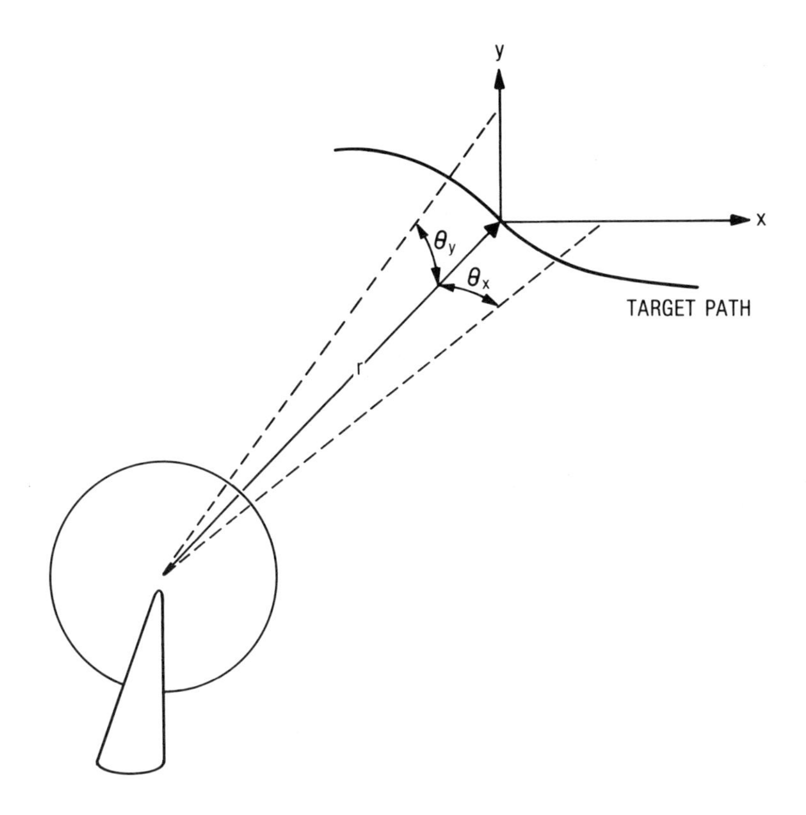

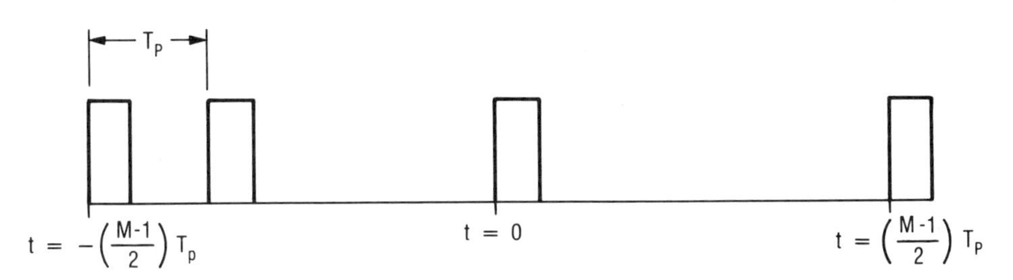

Figure 7.5 Typical tracking scenario and waveform.

The estimates $\hat{s}$ and $\hat{v}$ of s and v are given by [29]

$$\begin{bmatrix} \hat{s} \\ \hat{v} \end{bmatrix} = \begin{bmatrix} \dfrac{1}{M} & 0 \\ 0 & \dfrac{12}{T_{\mathrm{p}}^2(M+1)M(M-1)} \end{bmatrix} \begin{bmatrix} \displaystyle\sum_{k=-(M-1)/2}^{(M-1)/2} z(t_k) \\ T_{\mathrm{p}} \displaystyle\sum_{k=-(M-1)/2}^{(M-1)/2} kz(t_k) \end{bmatrix} \tag{7.77}$$

In order to determine the accuracy of these estimates the covariance matrix $\mathbf{Q}_2$ of $\hat{s}$ and $\hat{v}$ is needed, which is [29]

$$\mathbf{Q}_2 = \begin{bmatrix} \sigma_s^2 & 0 \\ 0 & \sigma_v^2 \end{bmatrix} = \sigma^2 \begin{bmatrix} \dfrac{1}{M} & 0 \\ 0 & \dfrac{12}{T_p^2(M+1)M(M-1)} \end{bmatrix} \tag{7.78}$$

Thus $\mathbf{Q}_2$ is the coefficient matrix in (7.77) multiplied by σ^2. To make use of this result note that at the end of the observation time $t_M = \frac{1}{2}(M-1)T_p$ the target position is

$$z(t_M) = s + vt_M + \tfrac{1}{2}at_M^2 \tag{7.79}$$

and the variance σ_z^2 of z is

$$\sigma_z^2 = E[(z(t_M) - \bar{z})^2] = E[s - \bar{s} + t_M(v - \bar{v}) + \tfrac{1}{2}t_M^2(a - \bar{a})]^2 \tag{7.80}$$

But $a \equiv \bar{a}$ since a is known and deterministic. Hence

$$\sigma_z^2 = \sigma_s^2 + \sigma_v^2 t_M^2 = \sigma_s^2 + \sigma_v^2 \left(\frac{M-1}{2}\right)^2 T_p^2 \tag{7.81}$$

where the cross terms between s and v vanish since by (7.78) the errors in s and v are uncorrelated. Referring to (7.78) the standard deviation of z is therefore

$$\sigma_z = \sigma\left[\frac{1}{M}\left(1 + \frac{3(M-1)}{M+1}\right)\right]^{1/2} \tag{7.82}$$

which is the standard deviation of the position measurement at the end of the tracking interval.

The particular coordinate to which (7.82) applies is identified by the appropriate value of σ. As was discussed in connection with (7.65) the general form of σ is $\sigma = k\,\Delta(2E/N_0)^{1/2}$ where Δ is the sensor-system measurement resolution for that coordinate. With regard to resolution, for range, from (7.35) $k\Delta = \Delta r/1.8$. For x and y the measurement resolution Δ in the x and y directions is $r\delta\theta_x$ and $r\delta\theta_y$ where $\delta\theta_x$ and $\delta\theta_y$ are the respective sensor beamwidths and r is the target range. For a sonar system employing a linear array there would be only one cross-range measurement direction, which would be nominally horizontal, and $\delta\theta \sim \lambda/L$ where λ is the signal wavelength and L the array length. For a radar employing a parabolic reflector, or a laser radar, the beam is circular and the symmetrical angular resolution $\delta\theta$ is nominally $4\lambda/\pi D$ where D is the optical aperture diameter or the antenna diameter (see Chapter 11); for these cases $k \sim 1/1.8$ as before.

Note that as $M \to \infty$ the term in brackets in (7.82) approaches $2/M$ so that,

$$\lim_{M\to\infty} \sigma_z \approx \frac{2k\Delta}{\sqrt{2ME/N_0}} \tag{7.83}$$

and for large M the process of observing the target during its track can be thought of simply as a means of increasing the effective signal-to-noise ratio by integrating the responses to M pulses. The factor of 2 in the numerator occurs because there are two unknown parameters. As will be seen shortly, if the acceleration is also unknown the factor in the numerator is 3.

Now suppose we wish to predict the target position at some time t_p after the end of the tracking interval T. It is only necessary to substitute $t_p + t_M$ for t_M in (7.79) and (7.82) becomes

$$\sigma_z \approx \frac{k\Delta}{[(2E/N_0)^{1/2}]} \left\{ \frac{1}{M} \left[1 + \frac{12\left(t_p + \frac{M-1}{2} T_p\right)^2}{T_p^2 (M+1)(M-1)} \right] \right\}^{1/2} \tag{7.84}$$

But if M is not too small $(M+1)T_p \sim (M-1)T_p \sim T$ where T is the total observation time. Hence (7.84) can be written

$$\sigma_z \approx \frac{k\Delta}{(2ME/N_0)^{1/2}} \left[1 + 12\left(\frac{t_p}{T} + \frac{1}{2}\right)^2 \right]^{1/2} \tag{7.85}$$

If $t_p \ll T$ this reduces to $\sigma_z \approx 2k\Delta/(2\mathrm{ME}/\mathrm{N}_0)^{1/2}$ as before. On the other hand, if $t_p \gg T$

$$\sigma_z \approx \frac{k\Delta}{(2ME/N_0)^{1/2}} \sqrt{12}\, \frac{t_p}{T} \tag{7.86}$$

and the error increases linearly with t_p/T. In all cases, E/N_0 is improved by a factor of M.

In this discussion only position measurements have been considered, and velocity and acceleration estimates have therefore been based on rates of change of target location. As is discussed in Chapter 8 however, range rate can also be measured by measuring the Doppler shift of the transmitted signal frequency. Suppose there is available both an estimate of range rate via Doppler shift, denoted as d, with variance σ_d^2, and an estimate of range rate, as above, by means of position measurements only, denoted as v with variance σ_v^2. Let V denote the estimate of range rate based on these two independent measurements. In this case V will be a weighted sum of the form $V = Ad + Bv$ where $A + B = 1$, and it can be shown that (see Exercise 7.13) the variance σ_V^2 of V is minimized by choosing A and B such that

$$V = \frac{\sigma_v^2}{\sigma_v^2 + \sigma_d^2}\, d + \frac{\sigma_d^2}{\sigma_v^2 + \sigma_d^2}\, v \tag{7.87}$$

and that in this case σ_V^2 is given by

$$\sigma_V^2 = \frac{\sigma_v^2 \sigma_d^2}{\sigma_v^2 + \sigma_d^2} \tag{7.88}$$

Equation (7.87) expresses the notion that the measurement with the smallest variance should get the greatest weight.

If Doppler measurements are available the foregoing results are modified by substituting $\sigma_v^2\sigma_d^2/(\sigma_v^2 + \sigma_d^2)$ for σ_v^2 in (7.81) and it can be shown (see Exercise 7.14) that (7.82) and (7.84) become: for measurement of range r at the end of the tracking interval

$$\sigma_r = \frac{\sigma}{\sqrt{M}}\left[1 + \frac{3(M-1)^2 T_p^2}{12\dfrac{\sigma^2}{\sigma_d^2} + T_p^2(M-1)M(M+1)}\right]^{1/2} \tag{7.89}$$

and for the predicted range t_p seconds after the observation interval

$$\sigma_r = \frac{\sigma}{\sqrt{M}}\left[1 + \frac{12\left[t_p + \left(\dfrac{M-1}{2}\right)T_p\right]^2}{12\dfrac{\sigma^2}{\sigma_d^2} + T_p^2(M-1)M(M+1)}\right]^{1/2} \tag{7.90}$$

where for a rectangular pulse spectrum, $\sigma = (c/3.6B)\sqrt{2E/N_0}$ where c is the speed of signal propagation. This result of course does not apply to the cross-range coordinates.

To summarize, s and v are estimated using (7.77) and the target position at future times is then determined using (7.75) where the time is measured from the center of the tracking interval. The position error at the end of the tracking interval is given by (7.82) and for later times the predicted-position error is given by (7.84) et seq; t_p here is measured from the end of the tracking interval, not from $t = 0$. If velocity measurements per se are also of interest the measurement error at the end of the tracking interval is $\sigma/T_p\{12/[(M+1)M(M-1)]\}^{1/2}$ where for range rate $\sigma = \Delta r/1.8$, and for cross-range motion $\sigma = 4\lambda r/1.8\pi D$ for a circular aperture of diameter D, where r is the range.

It is a straightforward matter to extend these results to the case where the acceleration is also unknown. From [29] the covariance matrix $\mathbf{Q}_3$ is

$$\mathbf{Q}_3 = \begin{bmatrix} \sigma_s^2 & 0 & \sigma_{sa}^2 \\ 0 & \sigma_v^2 & 0 \\ \sigma_{sa}^2 & 0 & \sigma_a \end{bmatrix}$$

$$= \sigma^2 \begin{bmatrix} \dfrac{3}{4}\dfrac{3M^2-7}{(M-2)M(M+2)} & 0 & \dfrac{-30}{T_p^2(M-2)M(M+2)} \\ 0 & \dfrac{12}{T_p^2(M-1)M(M+2)} & 0 \\ \dfrac{-30}{T_p^2(M-2)M(M+2)} & 0 & \dfrac{720}{T_p^4(M-2)(M-1)M(M+1)(M+4)} \end{bmatrix} \tag{7.91}$$

which shows that in this case the position and acceleration measurements are correlated. The estimates $\hat{s}$, $\hat{v}$ and $\hat{a}$ of s, v and a are [29]

$$\begin{bmatrix} \hat{s} \\ \hat{v} \\ \hat{a} \end{bmatrix} = \frac{\mathbf{Q}_3}{\sigma^2} \begin{bmatrix} \displaystyle\sum_{k=-(M-1)/2}^{(M-1)/2} z(t_k) \\ \displaystyle T_\mathrm{p} \sum_{k=-(M-1)/2}^{(M-1)/2} k z(t_k) \\ \displaystyle \frac{T_\mathrm{p}^2}{2} \sum_{k=-(M-1)/2}^{(M-1)/2} k^2 z(t_k) \end{bmatrix} \tag{7.92}$$

By following the same procedure used to obtain (7.82) the standard deviation of the position measurement at the end of the tracking interval is

$$\sigma_z = \frac{\sigma}{\sqrt{M}} \left[\frac{-3(7M^2 - 20M + 17)}{4(M-2)(M+2)} + \frac{3(M-1)}{(M+1)} + \frac{45(M-1)^3}{4(M-2)(M+1)(M+2)} \right]^{1/2} \tag{7.93}$$

which, as discussed above approaches $3\sigma/\sqrt{M}$ as $M \to \infty$. All the foregoing results obtained for the two-dimensional case can be obtained for unknown a using (7.91), (7.92) and (7.93).

EXERCISES FOR CHAPTER 7

7.1 Show that if $s(t) = \mathrm{Re}[\psi(t)e^{i2\pi f_0 t}]$, then $z_s(t) = s(t) + i\hat{s}(t)$ is equal to $\psi(t)e^{i2\pi f_0 t}$ only if $\psi(t)$ is band-limited to $\pm B/2$ with $B < 2f_0$. Use the result of (4.40), that if $a(t)$ is band-limited as above then the Hilbert transform of $a(t)\cos 2\pi f_0 t$ is $a(t)\sin 2\pi_0 t$.

7.2 Show that for any two analytic signals $z_a(t)$ and $z_b(t)$ that $z_a^*(t) * z_b(t) = 0$ and $z_a(t) * z_b(-t) = 0$.

7.3 Verify equations (7.6) through (7.12).

7.4 Show that the approximation of (6.4) is equivalent to the band-limiting assumption used in obtaining (7.14).

7.5 The output of a filter matched in amplitude only to a signal scattered from a target at range $r = ct_0/2$ can be written as

$$\int P_\mathrm{T}(\tau - t_0) P_\mathrm{T}(T - t + \tau) \cos 2\pi f_0(\tau - t_0) \cos[2\pi f_0(T - t - \tau) + \theta]$$

Show that the response of a square-law-detector to this input is independent of θ, but does not necessarily peak at $t = t_0 + T$.

7.6 Show that for a signal having a rectangular spectrum

$$H(f) = \begin{cases} A, & |f| \leq B/2 \\ 0, & \text{otherwise} \end{cases}$$

The definition (7.21) yields $\beta = 1.8B$.

7.7 Verify (7.22) and (7.23) using (7.18), (7.20) and (7.21).

7.8 Show that $f''(t_m) = 2E^2\beta^2$ by using (7.22), (7.23) and

$$\frac{d\zeta(t)}{dt} = \int_{-\infty}^{\infty} \psi(\tau - t_0)\, \frac{d\psi^*(T - t + \tau)\, d\tau}{dt}$$

$$= -\int_{-\infty}^{\infty} \psi(\tau - t_0)\, \frac{d}{d\tau}\, \psi^*(T - t + \tau)\, d\tau$$

$$= -\int_{-\infty}^{\infty} \psi(\tau - t_0)\psi'^*(T - t + \tau)\, d\tau$$

7.9 Show that if an estimator is not unbiased, then $E[(\hat{\theta} - \theta)^2]$ is not the variance of $\hat{\theta}$.

7.10 It is shown in the text that if y is a Gaussian random variable with known variance σ^2 and unknown mean μ, the maximum likelihood estimate of μ based on N observations is

$$\hat{\mu} = \frac{1}{N} \sum_{i=1}^{N} y_i$$

Show that if μ is known and σ^2 is unknown the (unbiased) MLE $\hat{\sigma}^2$ of σ^2 is

$$\hat{\sigma}^2 = \frac{1}{N} \sum_{i=1}^{N} (y_i - \mu)^2$$

and if both μ and σ^2 are unknown the MLE of σ^2 is

$$\hat{\sigma}^2 = \frac{1}{N-1} \sum_{i=1}^{N} (y_i - \hat{\mu})^2$$

where $\hat{\mu}$ is as above.

7.11 The range rate of a target is to be determined by measuring the Doppler shift. This is done by observing the shift in the peak of the spectrum of the transmitted pulse from the transmitted carrier frequency.

For a transmitted pulse $AP_{\mathrm{T}}(t) \cos 2\pi f_0 t$, calculate β of (7.57). Assume $f_0 T \gg 1$ and make the band-limited approximation. Repeat for a Gaussian pulse, $Ae^{-t^2/2\sigma_T^2} \cos 2\pi f_0 t$.

7.12 Referring to (7.12) $2E = \int |\psi(t)|^2\, dt = \int |\psi(t - t_0)|^2\, dt$ and therefore

$$\frac{d}{dt_0}(2E) = 0 = \frac{d}{dt_0}\int |\psi(t-t_0)|^2\,dt = -2\,\mathrm{Re}\int \psi'(t-t_0)\psi^*(t-t_0)\,dt$$

$$= -2\,\mathrm{Re}\int \psi'(t)\psi^*(t)\,dt = 0$$

Referring to (7.20) does this mean that it is always true that $\int f|\Psi(f)|^2\,df = 0$? Obviously not. Explain.

7.13 Verify (7.87) and (7.88). Hint: use Lagrange multipliers.

7.14 Verify (7.89) and (7.90).

8

WAVEFORM ANALYSIS, RANGE-DOPPLER RESOLUTION AND AMBIGUITY

From the results of the previous chapter it is clear that in order to measure the range to a target a sensor must transmit a pulse rather than a continuous signal, and in any application involving search, surveillance, or tracking, the sensor waveform would consist of a train of repeated pulses or repeated pulse bursts. This leads to the following problem. Consider a steady-state situation in which a sensor has been repeatedly transmitting pulses with a time separation between pulses of T_p seconds or, equivalently, a pulse repetition frequency (PRF) of $1/T_p$. Suppose after transmitting the nth pulse an echo is observed t_0 seconds later. This could be an echo of the nth pulse from a target at a range $r = ct_0/2$. On the other hand it could also be an echo of the $(n-1)$st pulse from a target at a range $(cT_p/2) + (ct_0/2)$ or from the $(n-2)$nd pulse from a target at range $(2cT_p/2) + (ct_0/2)$ and so on. Thus the use of a repeated waveform leads to an ambiguity in the measurement of range, and for a search volume of radius R it is necessary that T_p satisfy

$$\frac{cT_p}{2} \geq R \tag{8.1}$$

otherwise there will be an ambiguous range interval within the surveillance volume. For example, in a radar system, if $R = 1000$ km, then using $c = 3 \times 10^8$ m/s the PRF $1/T_p$ can be no greater than 150 pulses/sec.

There is also the possibility of ambiguity in the measurement of range rate. As will be shown, this can occur if the range rate is to be determined from measurement of the Doppler shift in the transmitted carrier frequency from an observation over a train of M pulses (on a pulse-by-pulse basis there is no such Doppler ambiguity for the sinusoidal (i.e. $BT \approx 1$) pulses under consideration here). However, as has been discussed in Chapter 7, target velocity can also be derived from tracking data for which there is no velocity ambiguity. Thus ambiguity in range rate is potentially less of a problem for sensing systems than ambiguity in range. In what follows we first deal with time-frequency analysis of periodically pulsed waveforms. The ambiguity characteristics of such waveforms are then considered, which are intimately connected with range-Doppler resolution capability.

8.1 WAVEFORM ANALYSIS

Let the transmitted waveform be a coherent train of pulses

$$w(t) = \sum_{n=0}^{M-1} \psi(t - nT_p) e^{i2\pi f_0 t} \tag{8.2}$$

where $\psi(t)$ is a complex low-pass pulse function of duration $T < T_p$ (see (7.5) and (7.6)). The transmitted signal spectrum $W_T(f)$ is then

$$\begin{aligned} W_T(f) &= \sum_{n=0}^{M-1} \int_{nT_p - T/2}^{nT_p + T/2} \psi(t - nT_p) e^{i2\pi f_0 t} e^{-i2\pi f t} \\ &= \sum_{n=0}^{M-1} e^{-i2\pi n T_p (f - f_0)} \int_{-T/2}^{T/2} \psi(t) e^{-i2\pi t (f - f_0)} \, dt \\ &= e^{-i\pi(M-1)T_p(f-f_0)} \frac{\sin \pi M T_p (f - f_0)}{\sin \pi T_p (f - f_0)} \Psi(f - f_0) \end{aligned} \tag{8.3}$$

where we have used

$$\sum_{n=0}^{M-1} p^n = \frac{1 - p^M}{1 - p} \tag{8.4}$$

and

$$\Psi(f) = \int_{-\infty}^{\infty} \psi(t) e^{-i2\pi f t} \, dt$$

The magnitude of $W_T(f)$ is shown in Figure 8.1a, which illustrates its essential features. (1) there is a spectral-line structure described by the function $\sin \pi M T_p (f - f_0) / \sin \pi T_p (f - f_0)$ which arises with the use of a repeated-pulse waveform. Because the waveform is coherent there is a peak at $f = f_0$. The frequency separation of the spectral lines is equal to the PRF, $1/T_p$. (2) the spectral resolution is determined by the width of the spectral

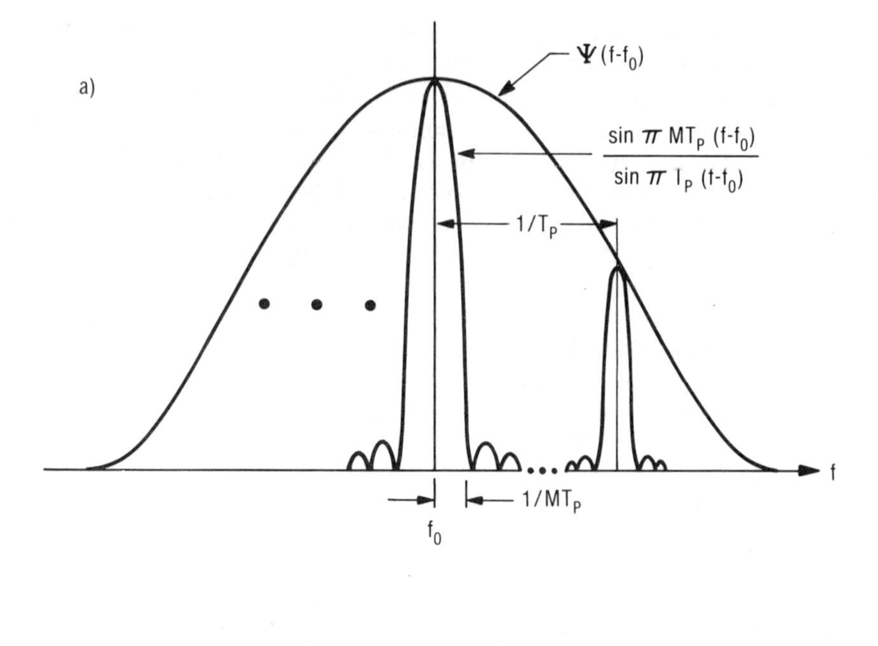

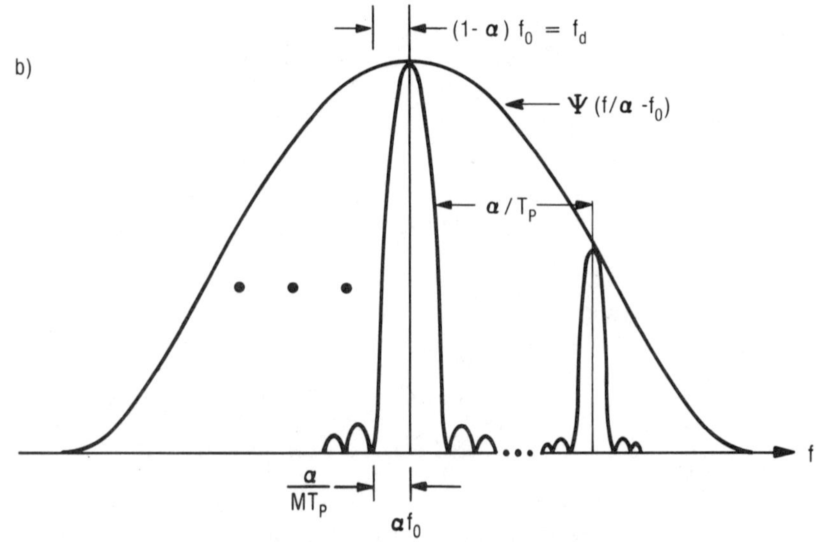

Figure 8.1 Magnitudes of spectra of transmitted and received coherent pulse burst.

lines. We say the resolution is nominally $1/MT_p$, being determined by the duration of the waveform $\sim MT_p$ not simply the number of pulses. Spectral peaks are separated by M sidelobe widths $1/MT_p$. (3) The overall shaping of the spectral lines is determined by the pulse spectrum $\Psi(f)$. Since the spectral lines of interest are always those in the vicinity of the peak, $f = f_0$, the actual pulse shape is less important in determining the essential spectral properties than the waveform duration and the PRF. Since the 3-dB spectral width of $\Psi(f)$ is nominally $\sim 1/T$ there will be $\sim T_p/T$ spectral lines contained within the frequency range of $\Psi(f-f_0)$.

A target approaching the sensor with constant range rate $v<0$ will at time t be located at some range $r(t)=r_0+vt$. Let transmission of $w(t)$ begin at $t=0$. As before, dispersion and all propagation effects on the signal amplitude and phase are ignored; if necessary, as in Ch. 7, all quantities can be renormalized so that E is the received signal energy. The received signal at time t is then $w(t-\tau(t))$ where $\tau(t)$ is the round-trip delay. To determine $\tau(t)$ observe that:

$$\tau(t)=\frac{2}{c}\,r(t-\tau(t)/2)=\frac{2}{c}\left[r_0+vt-\frac{v\tau(t)}{2}\right]$$

from which $\tau(t)=(2r_0+2vt)/(c+v)$ and the received signal is

$$w(t-\tau(t))=w\left(\left(\frac{c-v}{c+v}\right)t-\frac{2r_0}{c+v}\right)$$

Therefore (8.2) becomes

$$w(\alpha t-t_0)=\sum_{n=0}^{M-1}\psi(\alpha t-t_0-nT_p)e^{i2\pi f_0(\alpha t-t_0)} \tag{8.5}$$

where

$$\alpha=\frac{c-v}{c+v}\approx 1-\frac{2v}{c}$$

$$t_0=\frac{2r_0}{c+v} \tag{8.6}$$

and in the approximation for α terms of order v^2/c^2 have been ignored.

By applying the Fourier-transform operation of (8.3) to (8.5) it is found that the received signal spectrum $W_R(f)$ is

$$W_R(f)=\exp\left(-i2\pi\frac{f}{\alpha}t_0\right)exp\left[-i\pi(M-1)T_p\left(\frac{f}{\alpha}-f_0\right)\right]$$
$$\times\frac{\sin[\pi M(f/\alpha-f_0)T_p]}{\sin[\pi(f/\alpha-f_0)T_p]}\,\frac{\Psi(f/\alpha-f_0)}{\alpha} \tag{8.7}$$

which reduces to (8.3) for $t_0=0$, $\alpha=1$.

The magnitude of (8.7) is illustrated in Figure 8.1b. The following features are to be noted, all of which arise because the approaching target in producing the echo produces a time compression in the transmitted waveform. (1) The spectral peak has been shifted from f_0 to $\alpha f_0\approx(1-2v/c)f_0=f_0+f_d$ where $f_d=(-2v/c)f_0$ is the Doppler frequency. (2) Since each pulse has been time compressed to T/α, the spectral width of $\Psi(f)$ has been increased to α/T. (3) Since the separation between pulses has been reduced to T_p/α, the spectral resolution has been degraded to α/MT_p and the width between spectral peaks has been increased to α/T_p. If in the example the target were receding, then $\alpha<1$, all compressions become expansions, decreases become increases, and so on. Effect (1) is very important. Whether effects (2) and (3) may be ignored or not will depend on the particular application, as will be discussed.

The pulse train (8.2) is coherent[†] because the phase is contiguous from pulse to pulse over the entire waveform. One can think of (8.2) as a CW sinusoid of duration $(M-1)T_p$ with periodic interruptions in amplitude but not in phase. On the other hand, a noncoherent pulse train is of the form

$$w(t) = \sum_{n=0}^{M-1} \psi(t - nT_p) e^{i2\pi f_0(t-nT_p)} \tag{8.8}$$

which consists of a sum of M carrier pulses with no phase constraints. The phase relationship between successive pulses is essentially arbitrary, depending entirely on the choice of f_0 and T_p. It is left as an exercise to show that the transmitted signal spectrum in this case is

$$W_T(f) = e^{-i2\pi(M-1)T_p f} \frac{\sin \pi M T_p f}{\sin \pi T_p f} \Psi(f - f_0) \tag{8.9}$$

and the spectrum of the echo from a moving target is

$$\exp\left(-i2\pi \frac{f}{\alpha} t_0\right) \exp\left[-i\pi(M-1)T_p \frac{f}{\alpha}\right] \frac{\sin \pi M T_p f/\alpha}{\sin \pi T_p f/\alpha} \frac{1}{\alpha} \Psi\left(\frac{f}{\alpha} - f_0\right) \tag{8.10}$$

The spectral-line structure for the noncoherent transmitted and received waveforms depends only on the pulse spacing T_p, independent of the carrier frequency. Only the overall spectral shaping by $\Psi(f - f_0)$ contains any frequency information, as in fact it would if only a single pulse were transmitted. Also, aside from the factor $1/\alpha$, the spectral line structure remains unchanged and independent of the motion of the target, and the Doppler shift can therefore not be measured in this case since the information in the position of the peak of $\Psi(f - f_0)$ cannot be recovered because of the presence of $\sin \pi M f T_p / \sin \pi f T_p$. Therefore, if it were desired to measure Doppler shift coherency would be required.

The foregoing results illustrate important differences in the analysis of radar/laser radar and sonar waveforms. In the former case, even for a target traveling at orbital escape velocity of 7 km/s we have $\alpha = 1 - (2v/c) \approx 1.00005$ and all time-compression (or expansion) factors influencing resolution, signal bandwidth, etc. are essentially negligible. However, even for a range rate as small as 50 ft/s the Doppler shift $2vf_0/c$ for a typical radar frequency of 1.5×10^9 Hz is ~150 Hz which is easily measured. Therefore in radar and laser radar the effect of target motion can to a very good approximation be assumed to cause a change in carrier frequency only, and the observed frequency shifts are reasonably large.

[†] See page 135 for system coherency requirements.

In sonar the situation is quite different. For a target moving at 50 ft/s (30 knots) the value of $2v/c$ is 0.02. Although this would cause a bandwidth compression or expansion in the received signal of only a few percent, the change is many orders of magnitude larger than for radar or laser radar, and could be significant in applications requiring very sensitive phase measurements; this matter will be discussed further in connection with the ambiguity function. Also, for a typical value of $f_0 = 100$ Hz the Doppler shift in this case would be only 2 Hz. Furthermore, since the Doppler-velocity resolution Δv is $\Delta v = \lambda / 2T$, and since radar and laser radar wavelengths are many orders of magnitude smaller than for sonar, then to achieve the same Doppler-velocity resolution much longer signal durations are required for sonar than for radar.

8.2 RANGE-DOPPLER RESOLUTION AND AMBIGUITY—THE GENERALIZED AMBIGUITY FUNCTION

We now return to the question of range and Doppler ambiguity, the former of which has already been illustrated. In the latter case, referring to Figure 8.1, since the function $\sin \pi M(f/\alpha - f_0)T_p / \sin \pi(f/\alpha - f_0)T_p$ is periodic in frequency with period α / T_p, there is no means for determining whether or not a shift in the spectral-line component by an amount $\pm n\alpha / T_p$, $n = 0, 1, 2, \ldots$, may have occurred. Thus a target with a range rate v produces the same observable as a target with a range rate $v \pm n\alpha\lambda / 2T_p$. In physical terms, a target that moves in range a distance d in time T_p is ambiguous in range rate with a target that moves in range a distance $d \pm n\alpha\lambda/2$ in time T_p.

There are a number of ways of dealing with range and Doppler ambiguities. For range, it may be known a priori that there are no targets beyond the first ambiguous range interval given by (8.1), or we may not care, in which case all echoes are treated as if they are at ranges less than or equal to $cT_p/2$. In another approach, two different PRFs can be used which can be chosen such that in order to satisfy both solutions the unambiguous range interval extends out to where the presence of a target is not physically possible. Similarly, for range rate the system parameters can be chosen such that the ambiguous Doppler shifts can be ignored either because they are of no interest or impossible. Also, as noted, if range rate is derived from tracking data the range-rate ambiguity problem does not arise. In any case however, it is useful to be able to analyze a given proposed waveform in order to evaluate the range-Doppler ambiguities which may occur with its use. This is done by means of the ambiguity diagram which was evidently first proposed by Woodward [26].

The question of range-Doppler ambiguity is intimately connected with range-Doppler resolution capability. Let us begin therefore by first considering echoes $w(t - t_1)$ and $w(t - t_2)$ from two stationary targets at ranges $r_1 = ct_1/2$, $r_2 = ct_2/2$. For convenience let all distances be referred to r_1 so that we deal with $w(t)$ and $w(t+\tau)$, where $\tau = t_1 - t_2$ is the time separation between the echoes. In order to be able to resolve the two objects in range the composite return $w(t) + w(t+\tau)$ must exhibit two clearly defined peaks in time. Otherwise, if τ is too small, the composite will exhibit a single peak and the fact that two objects are present will not be discerned. Consider the mean-square composite echo

$$\int_{-\infty}^{\infty} |w(t) + w(t+\tau)|^2 \, dt = \int_{-\infty}^{\infty} |w(t)|^2 + \int_{-\infty}^{\infty} |w(t+\tau)|^2 + 2 \operatorname{Re} \int_{-\infty}^{\infty} w(t) w^*(t+\tau) \, dt \tag{8.11}$$

Clearly, if $w(t)$ and $w(t+\tau)$ do not overlap they will be completely resolved and the cross term will vanish. Therefore, a measure of the capability of the waveform to resolve the two targets is given by a correlation function $C_{\mathrm{R}}(\tau)$:

$$C_{\mathrm{R}}(\tau) = 2 \operatorname{Re} \int_{-\infty}^{\infty} w(t) w^*(t+\tau) \, dt = 2 \operatorname{Re} e^{-i2\pi f_0 \tau} \int_{-\infty}^{\infty} \psi_N(t) \psi_N^*(t+\tau) \, dt \tag{8.12}$$

where, referring to (8.2)

$$\psi_N(t) = \sum_{n=0}^{N-1} \psi(t - nT_{\mathrm{p}}) \tag{8.13}$$

By the Schwarz inequality (5.66), (8.12) will of course be maximum for $\tau = 0$, and the range-resolution capability of the waveform will be determined by the extent to which $C_{\mathrm{R}}(\tau)$ is negligible for $\tau \neq 0$. There is a parallel here between this discussion and that of Section 7.3.1 regarding the extent to which phase information can be used to measure range. Here we are interested in measuring range without using the phase information in the signal and shall therefore be concerned with evaluating the modulus $|C_{\mathrm{R}}(\tau)|$ of $C_{\mathrm{R}}(\tau)$, and shall ignore the phase term in (8.12).

As an example, let $\psi(t)$ be a rectangular pulse

$$\psi(t) = \begin{cases} 1, & -T/2 \leq t \leq T/2 \\ 0, & \text{otherwise} \end{cases} \tag{8.14}$$

Then $\psi_N(t)$ is as illustrated in Figure 8.2a, and $|C_{\mathrm{R}}(\tau)|$ is as illustrated in Figure 8.2b. By the foregoing discussion, for those time separations τ where $|C(\tau)|$ is negligible two targets can be resolved in time, or equivalently in range. Thus Figure 8.2b demonstrates that targets separated by T_{p} seconds are indistinguishable or, equivalently, as outlined in the foregoing discussion, the response to a target at range r is ambiguous with the response to a target at $r + ncT_{\mathrm{p}}/2$, $n = 1, 2, 3, \ldots$. Figure 8.2 also illustrates that the

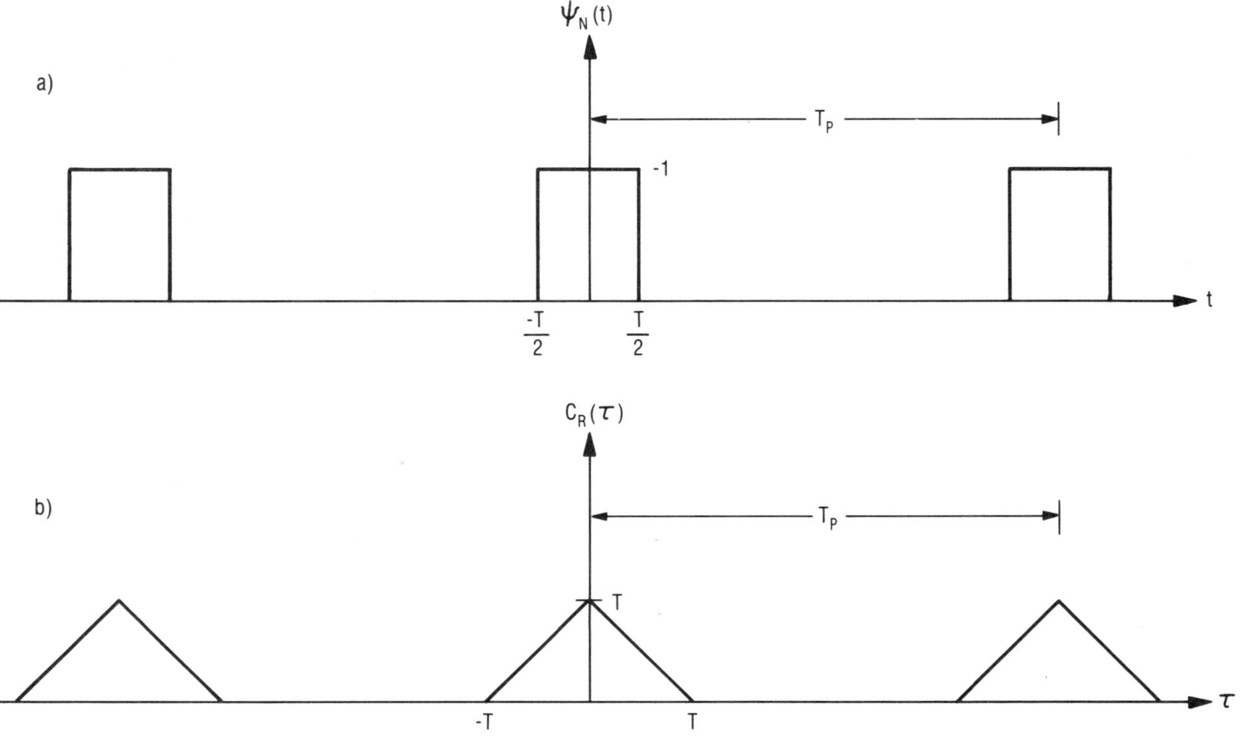

Figure 8.2 Train of rectangular pulses and associated correlation function.

shorter the pulse the greater the range resolution and the greater the unambiguous intervals, since the peaks of $|C_R(\tau)|$ thereby become narrower. Actually, it is the signal bandwidth rather than transmitted pulse duration that more fundamentally determines resolution capability, as will be discussed in Chapter 9. For purely sinusoidal waveforms however, for which the time-bandwidth product BT is essentially unity, there is no such distinction since $B \sim 1/T$.

Let us now consider the alternative situation in which two targets at the same point in space are moving at two different ranges rates; equivalently, let one target be stationary and let the other move with a relative range rate v. Then setting $\tau = 0$ and following exactly the same argument associated with equation (8.11) et seq. the correlation function of interest is

$$\int_{-\infty}^{\infty} w(t) w^*(\alpha t)\, dt \tag{8.15}$$

In this case, referring to (8.2) and (8.13),

$$\begin{aligned} C_D(f_d) &= \int_{-\infty}^{\infty} \psi_N(t) \psi_N^*(\alpha t) e^{i2\pi f_0 (1-\alpha) t}\, dt \\ &= \int_{-\infty}^{\infty} \psi_N(t) \psi_N^*(\alpha t) e^{-i2\pi f_d t}\, dt \end{aligned} \tag{8.16}$$

which is the Fourier transform of the product of two functions evaluated at $f = (1-\alpha) f_0 = -f_d$. By the complex convolution theorem (see Ex. 8.2) this is

$$\begin{aligned} C_D(f_d) &= \int_{-\infty}^{\infty} \Psi_N(f + f_d) \Psi_N^*\left(\frac{f}{\alpha}\right) \frac{df}{\alpha} \\ &= \int_{-\infty}^{\infty} \Psi_N(\alpha f + f_d) \Psi_N^*(f)\, df \end{aligned} \tag{8.17}$$

where

$$\Psi_N(f) = \int_{-\infty}^{\infty} \psi_N(t) e^{-i2\pi f t}\, dt$$

The Doppler information is contained in the frequency shift f_d. The question arises concerning the conditions under which the approximation $\alpha = 1$ is valid, since this greatly facilitates calculation of (8.17). It is sometimes stated that this approximation is valid when the fractional bandwidth B/f_0 is small (i.e. many cycles per pulse). This is clearly irrelevant since $\psi(t)$ is a low-pass function and (8.17) is independent of f_0. The important factor is v/c. For typical radar and laser-radar applications the approximation is valid since it has been shown that α is negligibly different from unity for all cases of current practical interest. For sonar, if $\Psi(f)$ is purely real the approximation is also valid because in this case (8.17) amounts to a convolution of spectral magnitudes such as Figures 8.1a,b and the difference in line width by a factor α which is typically 1.02 can be ignored. In general, however, $\Psi(f)$ will be complex and one must be concerned with the effect of α on the phases. Therefore, for sonar applications one must be careful about making this approximation.

Finally, the effects of simultaneous range-Doppler ambiguity can be incorporated into a single normalized function

$$\chi(\tau, f_{\mathrm{d}}) = \frac{\int_{-\infty}^{\infty} \psi_N(t)\psi_N^*(\alpha t + \tau)\exp(-i2\pi f_{\mathrm{d}} t)\, dt}{\int_{-\infty}^{\infty} \psi_N(t)\psi^*(\alpha t)\, dt}$$

$$= \frac{\int_{-\infty}^{\infty} \Psi_N(\alpha f + f_{\mathrm{d}})\Psi_N^*(f)\exp(-i2\pi f\tau)\, df}{\int_{-\infty}^{\infty} \Psi(\alpha f)\Psi^*(f)\, df} \tag{8.18}$$

where, as discussed in connection with (8.12) the factor $\exp(-i2\pi f_0 \tau)$ outside the integral has been suppressed. With this normalization $\chi(0,0) = 1$. The interpretation of (8.18) is that two targets having range and Doppler separation τ and f_{d} are resolvable in range and Doppler depending on the relative magnitudes of $|\chi(\tau, f_{\mathrm{d}})|$ and $|\chi(0,0)| = 1$ over the two dimensional τ, f_{d} plane. The resolution conditions are summarized as follows:

$|\chi(\tau, f_{\mathrm{d}})| \ll 1 \rightarrow$ targets resolvable

$|\chi(\tau, f_{\mathrm{d}}| \approx 1 \rightarrow$ resolution difficult

$|\chi(\tau, f_{\mathrm{d}})| = 1 \rightarrow$ resolution impossible; range-Doppler ambiguity

The ambiguity function is usually found the literature expressed as in (8.18) with $\alpha = 1$. This is because it was originally applied to radar in situations for which $\alpha = 1$ was a very good approximation.† In sonar this approximation is not necessarily valid and the more general form of (8.18) may be required. It is left as an exercise to prove that for $\alpha = 1$

$$\int_{-\infty}^{\infty}\int_{-\infty}^{\infty} |\chi(\tau, f_{\mathrm{d}})|^2\, d\tau\, df_{\mathrm{d}} = 1 \tag{8.19}$$

which may be interpreted as a conservation of ambiguity, in the sense that $|\chi(\tau, f_{\mathrm{d}})|$ cannot be made everywhere small for any waveform. In its more generalized form in (8.18), the integral under $|\chi(\tau, f_{\mathrm{d}})|^2$ is a function of α (see Exercise 8.5).

In order to illustrate and summarize these various effects in a simple manner consider, as illustrated in Figure 8.3a, a Gaussian pulse,

$$\psi(t) = \left(\frac{2E}{\sqrt{\pi}T}\right)^{1/2} \exp(-t^2/2T^2) \tag{8.20}$$

whose standard deviation T is a measure of its length; the normalization is such that $\int |\psi|^2\, dt = 2E$. It is left as an exercise to show that the ambiguity

† The approximation is of course made *after* we let $f_d = (\alpha - 1) f_0$.

function $\chi(\tau, f_d)$ is Gaussian with $\sigma_\tau = (1 + \alpha^2)^{1/2}T$ and $\sigma_{f_d} = (1 + \alpha^2)^{1/2}/2\pi T$, as illustrated in Figure 8.3b. Thus the ambiguity diagram yields the expected result that, excluding large-time-bandwidth signals, as the pulse duration increases the capability for range resolution decreases and that for Doppler resolution increases.

Now, as illustrated in Figure 8.4a, let the waveform consist of M Gaussian pulses, each separated by T_p seconds such that $(M - 1)T_p = T$, and each standard deviation $\Delta \ll T$. It is clear from the foregoing discussion that the ambiguity diagram appears as in Figure 8.4b. The overall Doppler resolution remains unchanged because the overall waveform duration is

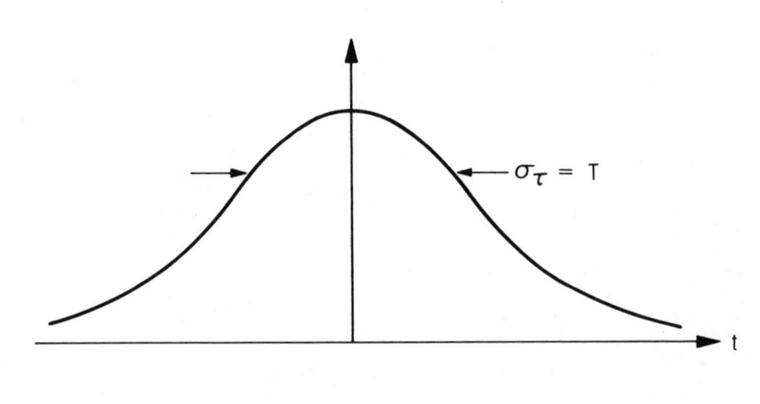

a) GAUSSIAN PULSE

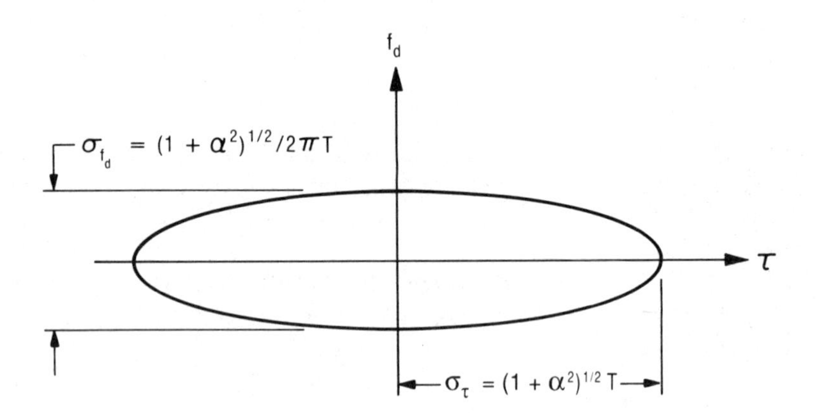

b) FOOTPRINT OF $|\chi(f_d, \tau)|$

Figure 8.3 Gaussian pulse and associated ambiguity function.

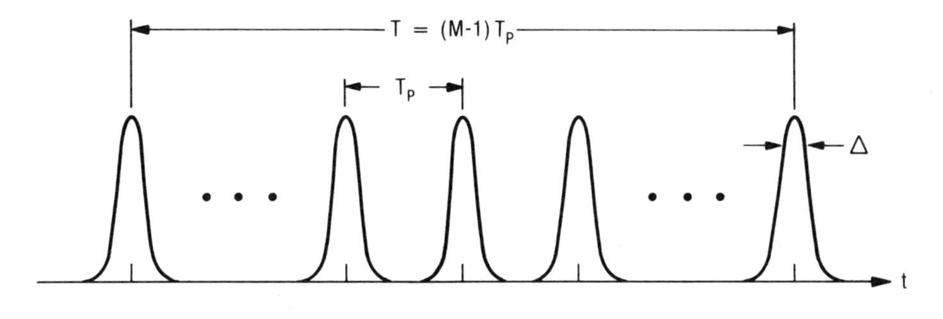

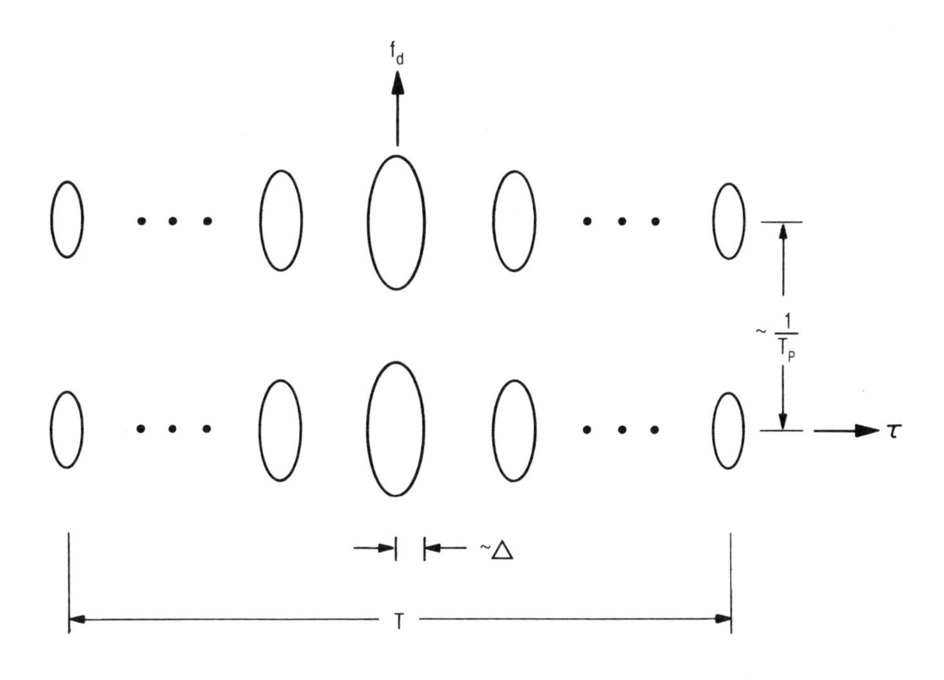

Figure 8.4 Gaussian pulse train and associated ambiguity function.

unchanged, and by transmitting M shorter pulses the range resolution has increased to nominally $c\Delta/2$ rather than $cT/2$. The price paid for this however is to introduce the range-Doppler ambiguities. To this illustration of the interplay between the various factors in waveform design must be added the constraint on the $\mathrm{PRF} = 1/T_\mathrm{p}$ imposed by the radius of the search volume in (8.1).

EXERCISES FOR CHAPTER 8

8.1 Verify (8.8), (8.9) and (8.10).

8.2 The complex convolution theorem states that $\int_{-\infty}^{\infty} x(t) * y^*(t)e^{-i2\pi ft}\, dt = X(f)Y^*(-f)$. Prove it, and show also that $\int_{-\infty}^{\infty} x(t)y^*(t)e^{-i2\pi ft}\, dt = \int_{-\infty}^{\infty} X(f_1 + f)Y^*(f_1)\, df_1$.

8.3 Show that the time and frequency representations in (8.18) are equivalent.

8.4 Prove that $\int_{-\infty}^{\infty} \int_{-\infty}^{\infty} |\chi(\tau, f_d)|^2\, d\tau\, df_d = 1$ for $\alpha = 1$.

8.5 Verify that the ambiguity function of the Gaussian pulse of (8.20) is also Gaussian with $\sigma_\tau = (1 + \alpha^2)^{1/2} T$ and $\sigma_{fd} = (1 + \alpha^2)^{1/2}/2\pi T$. Show that $\int_{-\infty}^{\infty} \int_{-\infty}^{\infty} |\chi(\tau, f_d)|^2\, d\tau\, df_d = \frac{1}{2}(1 + \alpha^2)$.

9

LARGE-TIME-BANDWIDTH WAVEFORMS

In all of the foregoing material only single-frequency sinusoidal pulses have been considered. These are signals for which the product of time duration T and bandwidth B is essentially unity. For such waveforms there is an inherent conflict between long-range detection and high range-resolution capability. Because the amplitude of an echo scattered from a target at range r decreases as r increases (see Chapter 11), then large values of r require large transmitted signal amplitudes in order to have sufficiently large values of E/N_0 for reliable detection and range estimation. But all active sensing systems have a limitation on the peak transmitted signal power, which imposes an upper limit on the transmitted signal amplitude. Of course the required value of E can also be obtained by maintaining the transmitted signal amplitude at some maximum value A and increasing the signal duration T. But since $B \sim 1/T$ and, as has been mentioned above and will be proved shortly, it is the signal bandwidth which is the fundamental parameter determining range-resolution capability, then achieving the required E by increasing T reduces B, thereby degrading range-resolution capability; hence, the conflict.

On the other hand, if B can be increased essentially independently of T there is no such conflict, and we have seen in Section 4.3 that there is no fundamental upper limit on the value of BT. As an example, suppose the peak transmitted-power limitation is $A^2/2$. Then for a given pulse duration T the transmitted energy E_T is $A^2T/2 = A^2/2B$ for $BT = 1$. Now suppose $BT = 100$. The transmitted energy is then $A^2T/2 = 100A^2/2B$ which is $100E_T$ for the same value of B, and therefore no degradation in range resolution. The limitation here is on the average power $100A^2/2BT_p$, where T_p is the inter-pulse spacing, and therefore the PRF, $1/T_p$, must be chosen appropriately. But this generally imposes less of a restriction than the peak-power limitation.

To show that B is the fundamental parameter which limits range resolution, refer to Chapter 8, (8.12), where it is shown that range-resolution capability is described by the magnitude of the correlation function $|C_R(\tau)|$ given by

$$|C_R(\tau)| = \left| \int_{-\infty}^{\infty} w(t)w^*(t+\tau)\, dt \right| \tag{9.1}$$

where τ is the relative time separation between two targets, corresponding to a separation in range of $c\tau/2$ for a rectangular pulse.

We recall that the more narrowly confined $|C_R(\tau)|$ is in the vicinity of its peak value, the greater the resolution capability. For purposes of illustration, let $w(t)$ be a single pulse rather than a pulse train. It might then be thought simply that the narrower the pulse $|w(t)|$ the narrower will be $C_R(\tau)$. But this is not necessarily true because $w(t)$ is complex and its phase also enters into the calculation in (9.1). On the other hand, using

$$W(f) = \int_{-\infty}^{\infty} w(t)e^{-i2\pi ft}\, dt$$

we have

$$|C_R(\tau) = \int_{-\infty}^{\infty} |W(f)|^2 e^{-i2\pi f\tau}\, df$$

which is the Fourier transform of $|W(f)|^2$ which is purely real and nonnegative. Now, the width of $|W(f)|^2$ directly (inversely) affects the width of $|C_R(\tau)|$. For example, let $|W(f)|^2$ be a unit-amplitude rectangle extending from $-B/2$ to $B/2$, in which case $|C_R(\tau)| = |\sin \pi B\tau / B\tau|$ which becomes narrower as B increases. Thus the bandwidth B of $w(t)$ is the fundamental parameter determining range resolution, which for certain waveforms can be increased essentially independently of the time duration of $w(t)$.

One of the earliest and probably most widely used large-BT signal is the linear FM (LFM) or Chirp waveform, a history of which is given in [30] and [31]. In addition to providing a solution to the long-range/high-resolution problem, the chirp signal is also a form of Doppler-invariant waveform. In producing an echo, a moving target alters the structure of the transmitted signal, and a filter which is matched to the transmitted signal may therefore be mismatched to its echo. For example, for $BT = 1$ waveforms, the carrier frequency of the echo from a target approaching at a constant range rate v is $f_0[1-(2v/c)] = f_0 + f_d$, and a filter matched to the transmitted signal may be translated out of the receiver passband and produce no response at all. In this case it would be necessary to employ a bank of filters each tuned to a different value of $f_0 \pm f_d$ over the entire range of possible Doppler shifts, and v would then be determined by noting which filter produces the maximum response. For the chirp waveform however, a filter matched to the transmitted signal is, within certain limits which will be derived below, also automatically matched to the echo from any constant velocity target with arbitrary range rate within the limits to be prescribed. Because, however, these limits may be restrictive in certain cases, particularly in sonar systems, an alternative large BT modulation scheme, hyperbolic frequency modulation, which is essentially free of such limitations, is also discussed.

9.1 CHIRP WAVEFORMS AND PULSE COMPRESSION

For a chirp waveform the analytic signal is

$$w(t) = \text{rect}\left(\frac{t}{T}\right) \exp\left[i2\pi\left(f_0 t + \frac{k}{2} t^2\right)\right] \tag{9.2}$$

where for simplicity in the notation which follows the signal is defined to extend over the interval $-T/2 \leq t \leq T/2$ as described by the rect function, defined as

$$\text{rect}\left(\frac{t}{T}\right) = \begin{cases} 1, & -T/2 \leq t \leq T/2 \\ 0, & \text{otherwise} \end{cases} \tag{9.3}$$

The instantaneous frequency in (9.2) is

$$f_I = \frac{d}{dt}\left(f_0 t + \frac{k}{2} t^2\right) = f_0 + kt \tag{9.4}$$

where, as illustrated in Figure 9.1, $k = B/T$ is the slope of the instantaneous frequency, which increases linearly with time; the frequency can of course

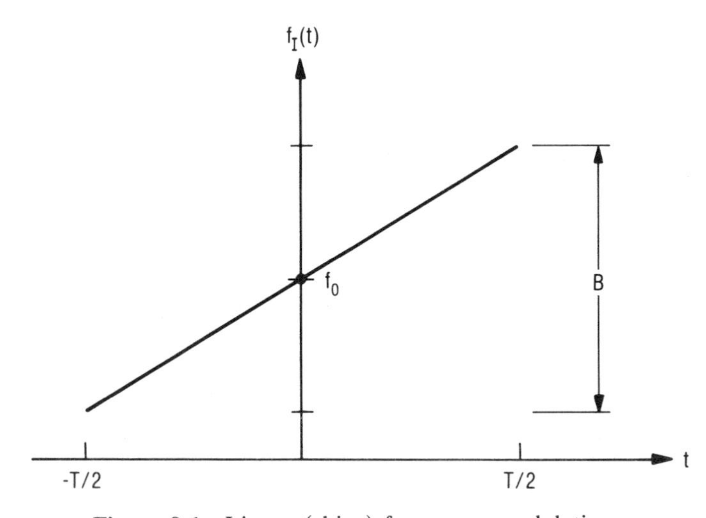

Figure 9.1 Linear (chirp) frequency modulation.

equally well be $f_I = f_0 - kt$, the former being commonly referred to as an upsweep and the latter as a downsweep.

We wish to determine the response of a filter matched to this signal. For purposes of this development it is simpler and clearer to deal with a noncausal matched filter, with impulse response $h(t) = Gw^*(-t)$, whose response to (9.2) is

$$
\begin{aligned}
x(t) &= G \int_{-\infty}^{\infty} w(\tau) w^*(\tau - t)\, dt \\
&= e^{i2\pi\left(f_0 t - \frac{kt^2}{2}\right)} G \int_{-\infty}^{\infty} \operatorname{rect}\left(\frac{\tau}{T}\right) \operatorname{rect}\left(\frac{\tau - t}{T}\right) e^{i2\pi k t \tau}\, dt
\end{aligned}
\tag{9.5}
$$

where G will be selected for unity gain. The integral in (9.5) is

$$
\begin{aligned}
&\int_{-T/2}^{T/2+t} e^{i2\pi k t \tau}\, d\tau\,, \qquad \text{for } -T \le t \le 0 \\
&\int_{t-T/2}^{T/2} e^{i2\pi k t \tau}\, d\tau \qquad \text{for } 0 \le t \le T
\end{aligned}
\tag{9.6}
$$

and by combining (9.5) and (9.6)

$$
x(t) = e^{i2\pi f_0 t} GT \operatorname{rect}\left(\frac{t}{2T}\right) \frac{\sin \pi B(|t| - t^2/T)}{\pi B|t|}
\tag{9.7}
$$

The matched-filter response is a single-frequency CW sinusoid with no phase modulation because the slope of the instantaneous frequency in the filter impulse response is opposite to that of the signal. It is left as exercises to verify (9.5), (9.6) and (9.7) and to show, as illustrated in Figure 9.2, that the maximum response occurs at $t = 0$, and that for large BT the zeroes in the vicinity of $t = 0$ occur at $t = \pm n/B$, $n = 1, 2, \ldots$, and the t^2/T term therefore has a small effect.

Now if the filter has unity gain, the energy at the output equals energy at the input, for which it is left as an exercise to show that $G = \sqrt{B/T}$. Thus the pulse amplitude has been raised by a factor $GT = \sqrt{BT}$. Also, for large BT, the matched filter has effectively compressed the pulse from a width T to a width $\sim 1/B$, thereby yielding a range resolution equal to the reciprocal of the transmitted pulse bandwidth, as expected. This technique is for this reason also referred to as pulse compression. The net effect is as if a pulse of duration $1/B$ and amplitude greater by a factor of $\sqrt{BT}$—which may have exceeded the peak power limitation—had been transmitted. We note the sidelobe structure in the filter response—the so-called range-time sidelobes, the first one of which is 13 dB below the peak of the main lobe for a rectangular pulse. If desired, these sidelobes can be reduced by ~15–20 dB by employing a nonrectangular transmitted pulse envelope, the price of

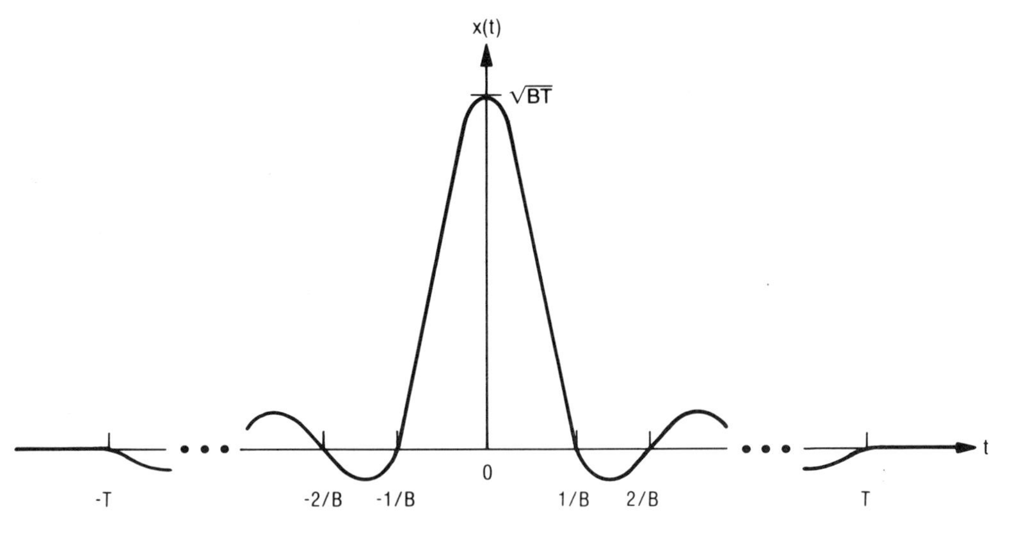

Figure 9.2 Response of filter matched to chirp signal, for $BT \gtrsim 50$.

which is to degrade the range resolution by broadening the main lobe somewhat [30, 31]. We do not pursue this further here.

Although the instantaneous frequency given by (9.4) sweeps over a range B, this does not necessarily give a true picture of the frequency content of the transmitted signal. In order to establish this we take the Fourier transform of (9.2) which is

$$W(f) = \int_{-\infty}^{\infty} \text{rect}\left(\frac{t}{T}\right) \exp\left[i2\pi\left(f_0 t + \frac{kt^2}{2}\right)\right] e^{-i2\pi ft}\, dt \tag{9.8}$$

After some manipulation this can be put into the form [30]:

$$W(f) = \exp\left[-i\pi \frac{(f-f_0)^2}{k}\right] \sqrt{\frac{T}{B}}\, [F(u_1) + F(u_2)] \tag{9.9}$$

where

$$F(u) = \int_0^u e^{i\pi x^2}\, dx \tag{9.10}$$

and

$$u_1 = \frac{1}{2}\sqrt{BT} + \frac{(f-f_0)}{\sqrt{k}}, \quad u_2 = \frac{1}{2}\sqrt{BT} - \frac{(f-f_0)}{\sqrt{k}} \tag{9.11}$$

Equation (9.10) is the Fresnel diffraction integral whose magnitude is shown in Figures 9.3a,b,c for three different values of BT; the vertical axes

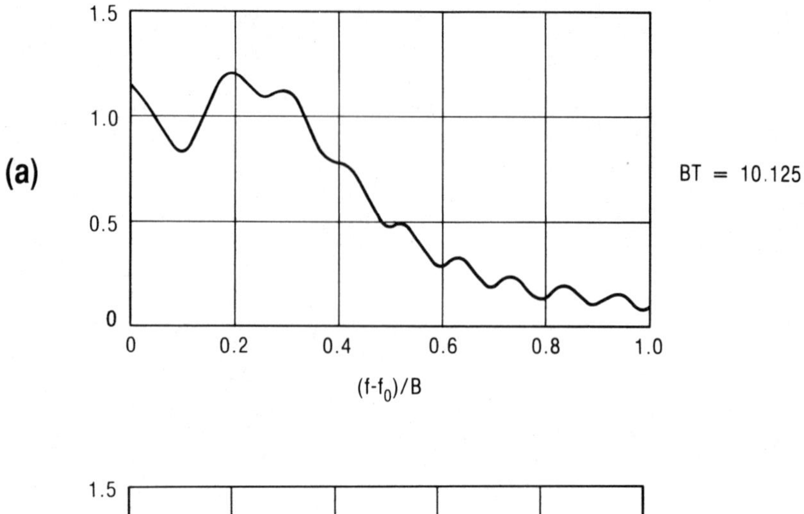

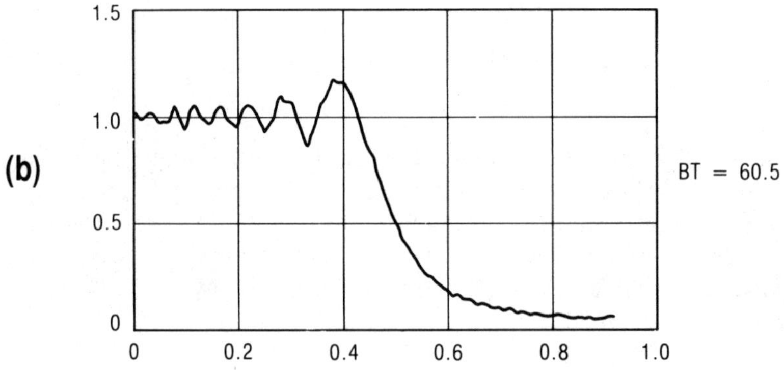

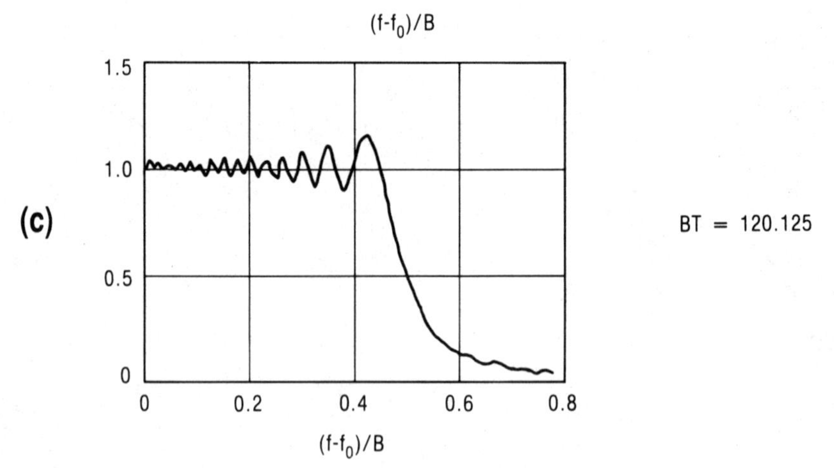

Figure 9.3 Magnitudes of spectrum of chirp signals for various values of BT (from [30]).

are in units of $1/\sqrt{k}$. Figures 9.3b and c are recognized as having the appearance of near-field, Fresnel-zone, knife-edge diffraction patterns which have the same mathematical form as chirp spectra. Also, the bandwidth is sensibly equal to the instantaneous frequency sweep B in these cases. For small values of BT however, $BT \approx 10$ in Figure 9.3a, the spectrum is very different and the signal bandwidth in this case is much less well defined, clearly not being simply equal to the frequency sweep B. Thus it is to be expected that the benefits to be gained with the use of Chirp pulses—of enabling large signal energy and high range resolution—require values of BT of, say, at least 50. This conclusion, based on examination of the signal spectrum, is consistent with the time domain analysis in which the zeroes of (9.7) are separated by $\Delta t \sim 1/B$ for large BT, as illustrated in Figure 9.2.

Equation (9.8) can also be evaluated to a sufficiently good approximation by the method of stationary phase, which will be of interest shortly. This is as follows. Consider the integral

$$I(\nu) = \int_a^b f(t) e^{i\nu h(t)}\, dt \tag{9.12}$$

For large ν the term $e^{i\nu h(t)}$ can oscillate rapidly, in which case $\int e^{i\nu h(t)}\, dt$ essentially vanishes because neighboring positive and negative areas cancel one another out. On the other hand, if at some point $a < c < b$, $h'(c) = 0$, the phase is stationary at this point, and $e^{i\nu h(t)}$ is therefore not oscillatory over the entire interval $[a, b]$ and $\int e^{i\nu h(t)}\, dt$ does not vanish. Hence, if $f(t)$ is generally slowly varying in comparison with $e^{i\nu h(t)}$, $I(\nu)$ can be calculated approximately by considering only those contributions to the integral in the vicinity of the stationary phase points. It can be shown [32] that for $\nu \gg 1$, $I(\nu)$ can in this way be approximated by

$$I(\nu) = \frac{\sqrt{2\pi} f(c) e^{i\nu h(c) \pm i\pi/4}}{[\pm \nu h''(c)]^{1/2}} + \mathcal{O}\left(\frac{1}{\nu^2}\right) \tag{9.13}$$

where + or − is used depending on whether $h''(c)$ is positive or negative. From the foregoing discussion the term $f(c)e^{i\nu h(c)}$ in (9.13) is to be expected; the remaining factor $(2\pi/\pm \nu h''(c))^{1/2}$ and $e^{\pm i\pi/4}$ follows from a contour integration.

To apply (9.13) to (9.8) the integral should be expressed in terms of dimensionless variables as

$$W(f) = T \int_{-\infty}^{\infty} \operatorname{rect}(u) \exp\left[i2\pi BT \left(v_0 u + \frac{u^2}{2} - uv \right) \right] du \tag{9.14}$$

where $u = t/T$, $v = f/B$, $v_0 = f_0/B$. The parameter ν in (9.12) is thus $2\pi BT$ and the approximation should therefore be very good for the range of values of BT of interest here.

In this case, $h(u) = v_0 u + \frac{1}{2}u^2 - uv$, and $dh(u)/du = 0$ yields $c = v - v_0$. Also, $h''(u) = 1$. Hence (9.13) yields

$$W(f) = T\sqrt{2\pi}\,\frac{\text{rect}(v - v_0)}{(2\pi BT)^{1/2}}\exp[-i\pi BT(v - v_0)^2 + i\pi/4]$$

$$= \sqrt{\frac{T}{B}}\exp\left[-i\pi\,\frac{(f - f_0)^2}{k} + \frac{i\pi}{4}\right]\text{rect}\left(\frac{f - f_0}{B}\right) \tag{9.15}$$

which may be compared with (9.9) et seq. and Figure 9.3b,c. The stationary-phase method has yielded the quadratic phase of $W(f)$ in (9.9) and also the essential features of the magnitude of the spectrum for large BT, excluding the Fresnel-integral oscillations in the magnitude.

Thus far it has been shown that the chirp signal provides high range resolution through the phenomenon of pulse compression and its frequency content has been determined as a function of BT. We now consider its Doppler-invariant properties.

9.2 DOPPLER INVARIANT PROPERTIES OF CHIRP WAVEFORMS

The foregoing matched filter response (9.7) is what would be observed for a motionless target at zero range. The effects of range delay and target motion on the filter response will now be explored. In this case as discussed in Chapter 8 in connection with (8.5), and ignoring as usual $1/r$ propagation attenuation etc., the echo is

$$w(t - \tau(t)) = w(\alpha t - t_0)$$

where $r(t) = r_0 + vt$, $\alpha \approx 1 - (2v/c)$ and $t_0 = (2r_0/(c + v)$. Hence the response of a filter matched to the transmitted signal is

$$x(t) = \sqrt{\frac{B}{T}}\int_{-\infty}^{\infty} w(\alpha\tau - t_0)w^*(\tau - t)\,d\tau$$

$$= \frac{1}{\alpha}\sqrt{\frac{B}{T}}\int_{-\infty}^{\infty} W\left(\frac{f}{\alpha}\right)\exp\left(-i2\pi\,\frac{f}{\alpha}\,t_0\right)W^*(f)e^{i2\pi ft}df \tag{9.16}$$

where $(1/\alpha)W(f/\alpha)\exp[-i2\pi(f/\alpha)t_0]$ is the spectrum of the echo. The spectrum $X(f)$ of the response of the matched filter to the echo is therefore,

$$X(f) = \frac{1}{\alpha}\sqrt{\frac{B}{T}}\,W\left(\frac{f}{\alpha}\right)e^{-i2\pi\frac{f}{\alpha}t_0}W^*(f)$$

and by applying the stationary-phase approximation (9.15) to $W(f/\alpha)$ and $W^*(f)$

$$X(f) = \frac{1}{\alpha}\sqrt{\frac{T}{B}}\,\mathrm{rect}\left(\frac{f/\alpha - f_0}{B}\right)\mathrm{rect}\left(\frac{f - f_0}{B}\right)$$
$$\times \exp\left[-i\,\frac{\pi}{k}\left(\frac{f}{\alpha} - f_0\right)^2\right]\exp\left[i\,\frac{\pi}{k}\,(f - f_0)^2\right]\exp\left(-i2\pi\,\frac{f}{\alpha}\,t_0\right) \tag{9.17}$$

The effect of the stationary phase approximation on the filter output may be illustrated by setting $\alpha = 1$, $t_0 = 0$ whence

$$x(t) = \int_{-\infty}^{\infty} X(f)e^{i2\pi ft}df = \sqrt{\frac{T}{B}}\int_{f_0 - B/2}^{f_0 + B/2} e^{i2\pi ft}\,df = e^{i2\pi f_0 t}\sqrt{BT}\,\frac{\sin \pi Bt}{\pi Bt} \tag{9.18}$$

Comparison with (9.7) shows that the stationary phase approximation leads to removal of the t^2 term in the argument of the $\sin x/x$ function which has been shown to be negligible for large BT. It also leads to a signal of infinite duration because of the bandlimiting properties of $\mathrm{rect}[(f - f_0)/B]$, which is an artifact of the approximation.

Equation (9.17) may be simplified for our purposes as follows. By using $a^2 - b^2 = (a + b)(a - b)$ the quadratic exponents can be written as

$$-\frac{i\pi}{k}\left[\left(\frac{f}{\alpha} - f_0\right)^2 - (f - f_0)^2\right]$$
$$= -\frac{i\pi}{k}\left[\left(f\left(\frac{1+\alpha}{\alpha}\right) - 2f_0\right)\left(f\left(\frac{1-\alpha}{\alpha}\right)\right)\right]$$
$$= -\frac{i\pi}{k}\left[f^2\left(\frac{1-\alpha^2}{\alpha^2}\right) - 2f_0 f\left(\frac{1-\alpha}{\alpha}\right)\right]$$

Now $(1 - \alpha^2)/\alpha^2 = 4v/c + \mathcal{O}(v^2/c^2)$ and $(1 - \alpha)/\alpha = 2v/c + \mathcal{O}(v^2/c^2)$ so ignoring terms of $\mathcal{O}(v^2/c^2)$ the exponent becomes, by completing the square

$$-\frac{i\pi}{k}\left[\frac{4v}{c}\,f^2 - \frac{4v}{c}\,ff_0\right]$$
$$= -i\pi\,\frac{4v}{c}\left[\frac{(f - f_0)^2}{k} + \frac{f_0 f}{k} - \frac{f_0^2}{k}\right]$$
$$= -i\pi\,\frac{4v}{c}\,\frac{(f - f_0)^2}{k} + i2\pi\,\frac{ff_{\mathrm{d}}}{k} + i\,\frac{4\pi v}{c}\,\frac{f_0^2}{k}$$

since $-4\pi v f_0/c = 2\pi f_{\mathrm{d}}$. Hence, ignoring the constant phase term $4\pi v f_0^2/ck$ and recognizing that if terms of $\mathcal{O}(v^2/c^2)$ are ignored we can write $f_d \approx f_{\mathrm{d}}/\alpha$

$$X(f) = e^{-i\pi\,\frac{4v}{c}\,\frac{(f-f_0)^2}{k}}\,\mathrm{rect}\left(\frac{f - f_0}{B}\right)$$
$$\times\,\frac{1}{\alpha}\sqrt{\frac{T}{B}}\,e^{i2\pi\,\frac{f}{\alpha}\left(\frac{f_{\mathrm{d}}}{k} - t_0\right)}\,\mathrm{rect}\left(\frac{f/\alpha - f_0}{B}\right) = A(f) \times B(f) \tag{9.19}$$

Thus, by the convolution theorem, the filter response $x(t)$ is a convolution of the inverse Fourier transforms $a(t)$ and $b(t)$ of $A(f)$ and $B(f)$. Taking the second one first

$$\begin{aligned} b(t) &= \left(\frac{T}{B}\right)^{1/2} \int_{-\infty}^{\infty} \text{rect}\left(\frac{f/\alpha - f_0}{B}\right) \exp\left[i2\pi \frac{f}{\alpha}\left(\frac{f_\text{d}}{k} - t_0\right)\right] e^{i2\pi ft} \frac{df}{\alpha} \\ &= \left(\frac{T}{B}\right)^{1/2} \exp\left[i2\pi f_0\left(\frac{f_\text{d}}{k} - t_0\right)\right] \int_{-\infty}^{\infty} \text{rect}\left(\frac{f}{B}\right) \\ &\quad \times \exp\left[i2\pi f\left(\frac{f_\text{d}}{k} - t_0\right)\right] \exp[i2\pi(f + f_0)\alpha t]\, df \qquad (9.20) \\ &= \sqrt{BT} \exp\left\{i2\pi f_0\left[\alpha t + \frac{f_\text{d}}{k} - t_0\right]\right\} \frac{\sin \pi B\left[\alpha t + \frac{f_\text{d}}{k} - t_0\right]}{\pi B\left[\alpha t + \frac{f_\text{d}}{k} - t_0\right]} \end{aligned}$$

Before dealing with the effect of the convolution with $a(t)$, which will result in a smearing of $b(t)$, let us examine $b(t)$, which is the essential undistorted filter response. This may be verified by setting $\alpha = 1$, $t_0 = 0$ in which case (9.20) reduces to (9.18), which is the matched-filter response under the stationary-phase approximation.

We observe in (9.20) that, whereas if the target were stationary the maximum filter response would occur at time $t = t_0$ (or $t = t_0 + T$ for a causal filter as discussed in Chapter 7, (7.15)), because of target motion the response occurs at $t = (1/\alpha)(t_0 - f_\text{d}/k)$. The time-compression/expansion factor α that affects B as well as t_0 is a very small effect here. The important points are that: (1) target motion is manifested as an apparent change in range—by an amount f_d/k—from what would otherwise be observed if the target were stationary, and (2) for purposes of target detection, within certain limits to be derived shortly, the same filter is matched to stationary as well as moving targets which, as discussed above, is not the case for $BT = 1$ waveforms. Item (2) of course simplifies the signal processing, but (1) means that the chirp pulse embodies an inherent range-Doppler ambiguity, in the sense that there are an infinite number of combinations of target positions and range rates which can be associated with any given observation of the time of occurrence of the maximum filter response. There are however methods for resolving this which are discussed in Section 9.3 in connection with the ambiguity function for large BT signals.

Thus, as discussed earlier, the chirp or LFM waveform is Doppler invariant in this sense. There is however a limit, which arises from the convolution of $b(t)$ with $a(t)$.

By the foregoing approximations, $A(f)$ in (9.19) is

$$A(f) = \exp\left[-i\frac{4\pi v}{c}\frac{(f - f_0)^2}{k}\right] \text{rect}\left(\frac{f - f_0}{B}\right) \qquad (9.21)$$

which, by referring to (9.15) is the spectrum of a chirp pulse of duration

$4vT/c$. For purposes of this discussion it is necessary to consider only the magnitude $|x(t)|$ of the filter output which is

$$\left|a(t) * b(t)\right| = \left|\sqrt{BT}\,\mathrm{rect}\left(\frac{t}{T'}\right)\exp\left[i2\pi\left(f_0 t - \frac{k'}{2}t^2\right)\right]\right.$$
$$\left. * \, e^{i2\pi f_0 \alpha t}\,\frac{\sin \pi B\left[\alpha t + \dfrac{f_\mathrm{d}}{k} - t_0\right]}{\pi B\left(\alpha t + \dfrac{f_\mathrm{d}}{k} - t_0\right)}\right| \tag{9.22}$$

where $k' = BT'$, $T' = |4vT/c|$, and $*$ denotes convolution. Equation (9.22) can be written as (see Exercise 9.7)

$$\left|\int_{-\infty}^{\infty} \sqrt{BT}\,\mathrm{rect}\left(\frac{t-\tau}{T'}\right)\frac{\sin \pi B\left[\alpha\tau + \dfrac{f_\mathrm{d}}{k} - t_0\right]}{\pi B\left[\alpha\tau + \dfrac{f_\mathrm{d}}{k} - t_0\right]}\right.$$
$$\left.\times \exp\left\{i2\pi\left[f_0(t-\tau) - \frac{k'}{2}(t-\tau)^2\right]\right\}\exp[i2\pi f_0(\alpha\tau + \alpha t - \alpha t)]\,d\tau\right|$$

$$= \sqrt{BT}\left|\exp\left[-i2\pi\left(f_\mathrm{d} t + \frac{k't^2}{2}\right)\right]\mathrm{rect}\left(\frac{t}{T'}\right) * \frac{\sin \pi B\left[\alpha t + \dfrac{f_\mathrm{d}}{k} - t_0\right]}{\pi B\left[\alpha t + \dfrac{f_\mathrm{d}}{k} - t_0\right]}\right| \tag{9.23}$$

Now the phase term $f_\mathrm{d} t + \frac{1}{2}k't^2$ is very slowly varying because $|f_\mathrm{d}| = |(1-\alpha)f_0| \ll f_0$ and also, $T' \ll T$. Therefore in (9.23) the phase will vary very little over the duration of $\mathrm{rect}(t/T')$ and to a good approximation

$$|x(t)| = |a(t) * b(t)| = \sqrt{BT}\left|\mathrm{rect}\left(\frac{t}{T'}\right) * \frac{\sin \pi B\left[\alpha t + \dfrac{f_\mathrm{d}}{k} - t_0\right]}{\pi B\left[\alpha t + \dfrac{f_\mathrm{d}}{k} - t_0\right]}\right| \tag{9.24}$$

Since the width of the convolution of two functions is nominally the sum of the widths of the two functions, the width Δ_x of the response $x(t)$ is

$$\Delta_x \approx \frac{1}{\alpha B} + \frac{1}{B}\frac{4v}{c}BT = \frac{1}{B}\left[\frac{1}{1+(2v/c)} + \frac{4v}{c}BT\right]$$
$$\approx \frac{1}{B}\left(1 + \frac{4v}{c}BT\right) \tag{9.25}$$

We could also write, approximately

$$\Delta_x = \frac{1}{B}\left[1+\left(\frac{4v}{c}BT\right)^2\right]^{1/2} \tag{9.26}$$

which would be exact if $a(t)$ and $b(t)$ were Gaussian. In either case, it has been shown that $1/B$ is the nominal width of the compressed pulse for a stationary target. Thus the convolution with $a(t)$ effectively reduces the bandwidth of the filter response by a factor $[1+(4v/c)BT]$, thereby degrading the range resolution by this amount.

In radar or laser radar for extreme values of v, say 7 km/s, values of $BT4v/c$ can be of the order of unity for relatively modest values of BT, in this case $BT \approx 8000$, and this effect must be considered. In many other applications of such systems however the effect can be ignored and target motion can be assumed to cause only a shift in carrier frequency from f_0 to $f_0 + f_d$. On the other hand, in sonar the effect is much more significant because of the much larger values of v/c. As an example, typical BT values can be ~1000, in which case, for $v = 30$ knots $BT4v/c \approx 40$ which would be unacceptable. In fact, typically, $BT4v/c \gtrsim 1$ for $v \gtrsim 1$ foot/s, which of course includes all targets of interest. In order to avoid this there is an alternative large BT modulation scheme [33] which can be employed, which will now be considered.

9.3 HYPERBOLIC FREQUENCY MODULATION (HFM)

It has been shown that a LFM signal $w(\alpha t)$ scattered from a moving target can be mismatched to a filter which is matched to the transmitted LFM signal $w(t)$. Referring to (9.16) which expresses the matched filter output, and to (9.2), this arises essentially because of the quadratic phase term in the integrand, $k(1-\alpha^2)\tau^2 \approx (B/T)(4v/c)\tau^2$. Thus, since the range of integration on τ is T, the mismatch will, as we have seen, be insignificant if $BT4v/c$ is small. The mismatch occurs because of the change in time scale, $w(t) \to w(\alpha t)$. However, if the phase modulation was such that $w(\alpha t) = w(t-t')$, that is, so that vehicle motion imparts an effective delay to the signal rather than a scale change, then only a phase shift would occur and there would be no degradation in range resolution or, in fact, matched-filter output power. This suggests a logarithmic phase modulation, because we can write

$$\log(a + b\alpha t) = \log \alpha(a + b(t-t')) = \log \alpha + \log(a + b(t-t'))$$

where $t' = a(\alpha - 1)/\alpha b$ and the residual constant phase shift is proportional to $\log \alpha$.

Now we note that if the instantaneous frequency is described by a hyperbola,

$$\frac{f_0}{1-(k/f_0)t} = f_0 + kt + \frac{k^2}{2f_0}\,t^2 + \cdots \tag{9.27}$$

which yields the chirp, linear frequency modulation to a first approximation, then the phase is

$$\int \frac{f_0}{1-(k/f_0)t}\,dt = \frac{-f_0^2}{k}\log\left(1-\frac{kt}{f_0}\right) \tag{9.28}$$

which is of the desired form. Thus, the transmitted signal is

$$w(t) = \mathrm{rect}\left(\frac{t}{T}\right)\exp\left[-i2\pi\frac{f_0^2}{k}\log\left(1-\frac{kt}{f_0}\right)\right] \tag{9.29}$$

Following the procedure of the preceding section, the transmitted signal spectrum is

$$W(f) = \int_{-\infty}^{\infty}\mathrm{rect}\left(\frac{t}{T}\right)\exp\left[-i2\pi\frac{f_0^2}{k}\log\left(1-\frac{kt}{f_0}\right)\right]e^{-i2\pi ft}\,dt \tag{9.30}$$

which using (9.13) is

$$W(f) \approx \sqrt{\frac{1}{k}}\,\mathrm{rect}\left(\frac{1-f_0/f}{B/f_0}\right)\frac{f_0}{f}\exp\left[-i2\pi\frac{f_0^2}{k}\log\left(\frac{f_0}{f}\right)\right]$$
$$\times\exp\left(-i2\pi\frac{ff_0}{k}\right)\exp\left(i2\pi\frac{f_0^2}{k}\right)\exp(+i\pi/4) \tag{9.31}$$

where

$$\mathrm{rect}\left(\frac{1-f_0/f}{B/f_0}\right) = \begin{cases} 1, & \dfrac{f_0}{1+(B/2f_0)} \le f \le \dfrac{f_0}{1-(B/2f_0)} \\ 0, & \text{otherwise} \end{cases} \tag{9.32}$$

and the Fourier transform of the filter output is—ignoring unity-gain considerations,

$$\frac{1}{\alpha}W\left(\frac{f}{\alpha}\right)\exp\left(-i2\pi\frac{ft_0}{\alpha}\right)W^*(f) = \frac{1}{k}\exp\left(-i2\pi\frac{f_0^2}{k}\log\alpha\right)$$
$$\times\,\mathrm{rect}\left(\frac{1-f_0/f}{B/f_0}\right)\mathrm{rect}\left(\frac{1-\alpha f_0/f}{B/f_0}\right)\left(\frac{f_0}{f}\right)^2\exp\left[-i2\pi f\left(\frac{t_0}{\alpha}+\frac{f_0}{k}\,\frac{(1-\alpha)}{\alpha}\right)\right] \tag{9.33}$$

The exponent contains only a term linear in f such as that in $B(f)$ in (9.19) and therefore there will be no degradation such as that brought about by the dispersive term $(f - f_0)^2$ in the phase of $A(f)$ in (9.19). Now since to a very good approximation

$$\operatorname{rect}\left(\frac{1 - f_0/f}{B/f_0}\right) \operatorname{rect}\left(\frac{1 - \alpha f_0/f}{B/f_0}\right) = \operatorname{rect}\left(\frac{1 - f_0/f}{B/f_0}\right)$$

the magnitude of the time response of the filter output is to a very good approximation

$$\begin{aligned} &\frac{1}{\alpha}\left|\int_{-\infty}^{\infty} W\left(\frac{f}{\alpha}\right) W^*(f) \exp\left[i2\pi f\left(t - \frac{t_0}{\alpha}\right)\right] df\right| \\ &\approx \frac{1}{k}\int_{f_0/(1+B/2f_0)}^{f_0/(1-B/2f_0)} \left(\frac{f_0}{f}\right)^2 \exp\left\{i2\pi f\left[t - \frac{1}{\alpha}\left(t_0 - \frac{f_d}{k}\right)\right]\right\} df \end{aligned} \tag{9.34}$$

We wish to determine (1) the time instant that the peak output occurs, and (2) the width of the main lobe of the time response which, as before, will be of the form $\sin x/x$. For these purposes it is not necessary to evaluate (9.34) explicitly. For (1), since the phase in the exponential is linear in f it is therefore equal to zero for all t satisfying

$$t = \frac{1}{\alpha}\left(t_0 - \frac{f_d}{k}\right) \tag{9.35}$$

The maximum value of (9.34) therefore occurs for this value of t.[†] Thus HFM is Doppler invariant, independent of $|4BTv/c|$, and exhibits the range-Doppler ambiguity as discussed for the chirp waveform.

For (2), since $(f_0/f)^2$ is slowly varying, (9.34) will to a very good approximation have zeroes for values of t for which there are an integral number of cycles in the range of integration

$$\frac{f_0}{1 - (B/2f_0)} - \frac{f_0}{1 + (B/2f_0)} = \frac{B}{1 - (B^2/4f_0^2)}$$

We therefore require

$$\left[t - \frac{1}{\alpha}\left(t_0 - \frac{f_d}{k}\right)\right] \frac{B}{1 - (B^2/4f_0^2)} = n \tag{9.36}$$

for $n = \pm 1, 2$, etc. This is satisfied by

$$t = \frac{1}{\alpha}\left(t_0 - \frac{f_d}{k}\right) + \frac{n}{B}\left(1 - \frac{B^2}{4f_0^2}\right) \tag{9.37}$$

[†] By a stationary-phase argument.

and the compressed pulse width is negligibly different from $1/B$, independent of $|4BTv/c|$ since we always have $B/f_0 \leq 1$. (If a more exact evaluation of possible degradation in range resolution is required the integral in (9.34) should be evaluated.) Thus HFM offers the possibility of avoiding the degradation in range resolution that can accompany the use of LFM waveforms in sonar applications, and also in radar and laser radar but generally less often.

9.4 AMBIGUITY FUNCTIONS FOR LARGE BT WAVEFORMS

The ambiguity function is (ignoring the normalizing factor)

$$|X(\alpha, \tau)| = \left| \int_{-\infty}^{\infty} w(t) w^*(\alpha t + \tau) \, dt \right| \tag{9.38}$$

which is only trivially different from the matched filter response (9.16); to make the correspondence exact, in (9.16) set $\tau = x$,[†] $t_0 = -\tau$ and $t = 0$. Hence, in discussing the properties of $|X(\alpha, \tau)|$, or $|X(f_d, \tau)|$ as the case may be, the results of Section 9.2 can be applied directly.

For the LFM pulse, the footprint of $|X(f_d, \tau)|$ is determined at once from (9.24) by the foregoing transformation; also we assume $(4v/c)BT \ll 1$ so that the convolution with rect function in (9.24) can be ignored. The essential features of the footprint in this case is illustrated in Figure 9.4a. It is seen that it is possible to simultaneously achieve a range resolution commensurate with the instantaneous frequency sweep B and a Doppler resolution commensurate with the transmitted pulse duration T. For example, if $BT = 1000$ then a range resolution can be achieved equivalent to a pulse duration $T/1000$ seconds, and at the same time a Doppler resolution can be achieved equivalent to $B/1000$ Hz. This may be compared with a CW, $BT = 1$, pulse of duration T where, as illustrated in Figure 9.4b, although a Doppler resolution $1/T$ can be achieved, the range resolution would be 1000 times as coarse.

The solid-line shape in Figure 9.4a corresponds to an instantaneous frequency with an upsweep, that is $f_I = f_0 + kt$. However an instantaneous frequency with a downsweep, $f_I = f_0 - kt$ is equally possible, for which the corresponding diagram is shown with a dotted line. From this it is clear that illumination of a target by an upsweep signal followed by a downsweep is one way to resolve the LFM range-Doppler ambiguity, provided of course the target range does not change appreciably during the interpulse spacing. More precisely, referring to (9.20), for an upsweep signal the peak filter response occurs at $t = (1/\alpha)(t_0 - f_d/k)$, and for a downsweep it occurs at $t = 1/\alpha(t_0 + f_\alpha/k)$. Thus, the arithmetic mean of the time instances of maximum response for two successive pulses with opposite sweeps yields t_0/α, and one-half of the difference yields $f_d/\alpha k$. Of course, as is discussed in

[†] Dummy variable.

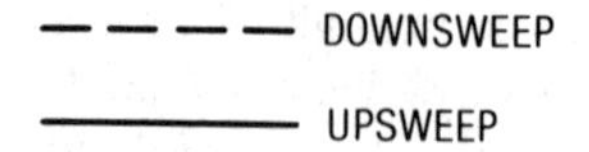

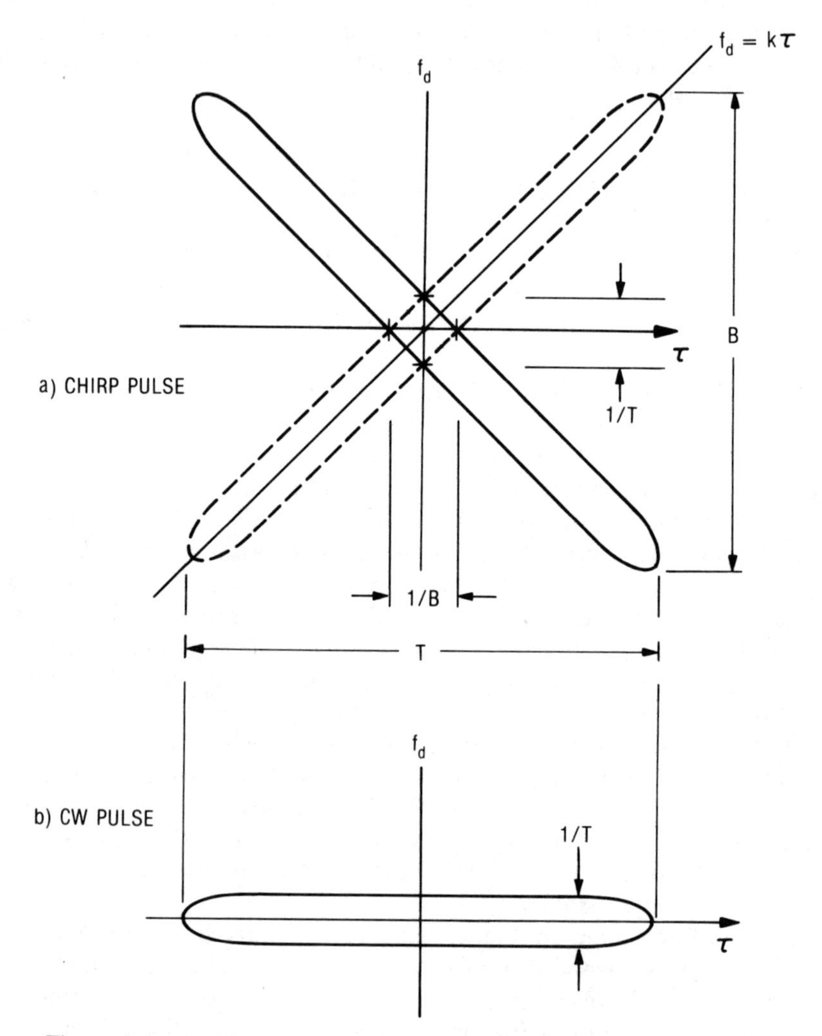

Figure 9.4 Ambiguity footprints for CW and LFM or HFM pulses.

the introduction to Chapter 8, if the range rate can be determined by means of tracking this problem is avoided. The slight error in the measurement caused by $1/\alpha$ is of second order.

If we write out $|X(\alpha, \tau)|$ exactly for the chirp signal, it is, ignoring terms of $\mathcal{O}(v^2/c^2)$

$$|X(\alpha, \tau)| = \left| \int_{-\infty}^{\infty} \text{rect}\left(\frac{t}{T}\right) \text{rect}\left(\frac{\alpha t + \tau}{T}\right) \times \exp\left[-i2\pi k \alpha t\left(\frac{f_d}{k} + \tau\right)\right] \exp\left[i\pi \frac{4kvt^2}{c}\right] dt \right| \tag{9.39}$$

The quadratic term in the exponential is $(4v/c)(Bt^2/T)$ as before. Hence, if $(4v/c)BT$ is not too large the quadratic exponential term can be ignored and the integral achieves its maximum value for $f_d = -k\tau$, yielding the footprint of $|X(\alpha_1 \tau)|$ given by (9.38), with reference to (9.24).

On the other hand, consider the ambiguity function for HFM which, referring to (9.29) is

$$|X(\alpha, \tau)| = \left| \int_{-\infty}^{\infty} \text{rect}\left(\frac{t}{T}\right) \text{rect}\left(\frac{\alpha t + \tau}{T}\right) \times \exp\left\{-i2\pi \frac{f_0^2}{k}\left[\log\left(1 - \frac{kt}{f_0}\right) - \log\left(1 - \frac{k(\alpha t + \tau)}{f_0}\right)\right]\right\} dt \right| \tag{9.40}$$

The exponential can be written as

$$\frac{2\pi f_0^2}{k}\left[\log\left(1 - \frac{kt}{f_0}\right) - \log\left(1 - \frac{k(\alpha t + \tau)}{f_0}\right)\right] = \frac{2\pi f_0^2}{k}\left[\log\left(1 - \frac{kt}{f_0}\right) - \log \alpha\left(\frac{f_0 - k\tau}{\alpha f_0} - \frac{kt}{f_0}\right)\right]$$

and if $(f_0 - k\tau)/\alpha f_0 = 1$, or $\tau = (1 - \alpha)f_0/k = -f_d/k$ the exponential term becomes a constant, $\exp[-i2\pi(f_0^2/k)\log \alpha]$ and the integral in (9.40) achieves its maximum value. For HFM the ambiguity diagram follows the contour $f_d = \pm k\tau$ exactly in *all* cases, and does not require $(4v/c)BT \ll 1$. Thus, consistent with its Doppler-invariant properties, HFM yields an ambiguity diagram which is also devoid of the errors that occur in the LFM ambiguity diagram for large $4vBT/c$. Again, this makes HFM especially useful for sonar.

9.5 CODED WAVEFORMS

For completeness we discuss briefly another large BT waveform, which is generally of the form

$$w(t) = \sum_{n=1}^{M} a_n \, \text{rect}[B(t - n/B)] e^{i2\pi f_0(t - n/B)} \tag{9.42}$$

where

$$\text{rect}(Bt) = \begin{cases} 1, & -\dfrac{1}{2B} \leq t \leq \dfrac{1}{2B} \\ 0, & \text{otherwise} \end{cases}$$

and, typically, the a_n are elements of a pseudorandom sequence of ± 1s; these are also referred to as pseudonoise (PN) sequences and sometimes also as pseudorandom noise (PRN) waveforms. Thus the signal is a sequence of carrier pulses, with energy determined by the length of the sequence and bandwidth determined by the rate at which the phase reversals in the carrier take place. Such signals are also used in Direct-Sequence spread spectrum communications [34].

Clearly, the BT products for such signals can be made arbitrarily large. The matched filter for such a signal consists essentially of a tapped delay line with tap weights equal to the a_n, as illustrated in Figure 9.5. If the sequence a_n is chosen properly, the summed output, which is the autocorrelation function of the sequence, will be very small until correlation occurs, at which point the response will reach a magnitude $\sim M$ in a time period $\sim 1/B$. Since $a_n^2 = 1$ for all n, at the instant of correlation the contents of the output signal register consists of a CW sinusoid of duration $M/B = T$, which can be processed to determine Doppler shift with a frequency resolution of $1/T$. Thus the coded waveform exhibits the same desirable high-resolution properties of LFM and HFM, which is as expected since the fundamental quantity for such schemes is BT, regardless of the particular modulation.

The correlation function shown in Figure 9.6 is for a maximal-length [34] pseudorandom sequence for which correlation functions having this shape are guaranteed. In nonmaximum-length sequences the tails of the correlation function will in general be larger and may exhibit a sidelobe structure which of course is not as desirable. Maximal-length sequences however can be easily decoded and jammed by an adversary and more complicated nonlinear codes are often employed, for which the structure of the correlation function must be studied.

Ideally, the ambiguity diagram of PN sequences possess very desirable characteristics which is a major reason for interest in their use with active sensors. Clearly, the most desirable form of ambiguity function is the thumbtack function illustrated in Figure 9.7, which is devoid of all the range-Doppler ambiguities characteristic of CW pulse trains, LFM and HFM. Ideally, such thumbtack ambiguity functions, which are essentially two-dimensional versions of the correlation function in Figure 9.6, are possible using coded waveforms.

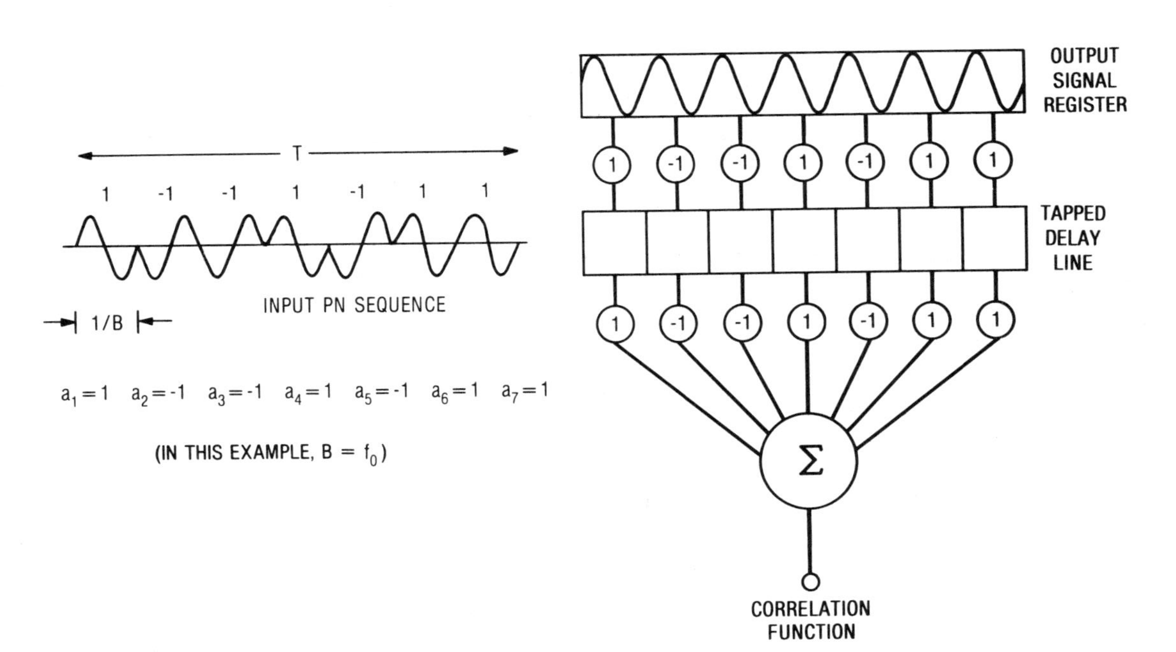

Figure 9.5 Receiver for PN sequence.

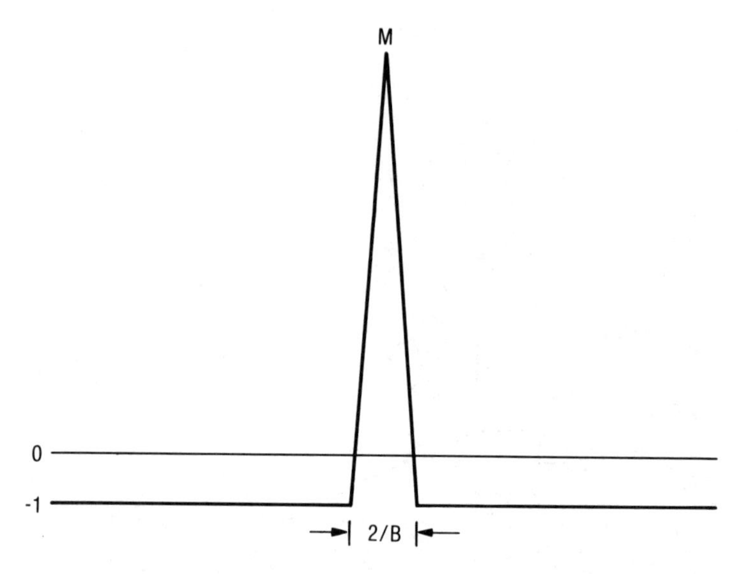

Figure 9.6 Autocorrelation function for *M*-pulse maximal-length PN sequence.

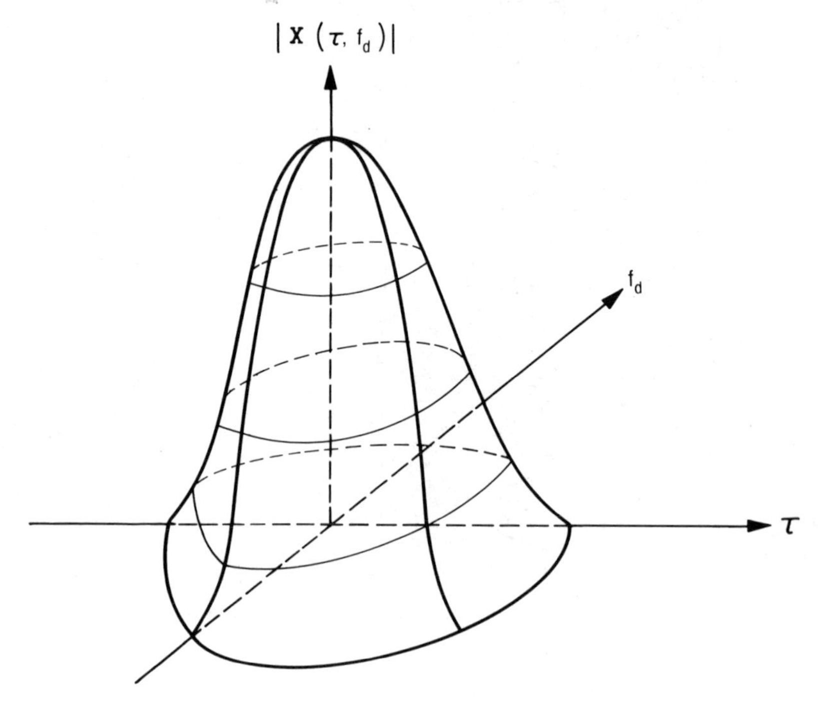

Figure 9.7 Thumbtack ambiguity function for PN sequence.

EXERCISES FOR CHAPTER 9

9.1 Verify (9.5), (9.6) and (9.7).

9.2 Show that for large BT, the zeros of (9.7) occur for $t = \pm n/B$, $n = 1, 2, \ldots$.

9.3 Verify (9.19) and (9.20).

9.4 Show that for large BT the energy $x(t)$ in (9.7) is G^2T^2/B; use equation (3.72) et seq. Therefore, for unity gain $G = \sqrt{B/T}$. Show that the pulse-compression operation can be represented as a system which increases the input amplitude by a factor of $\sqrt{BT}$ and reduces (divides) the pulse duration by a factor of BT.

9.5 Discuss the result of exercise 9.4 in terms of matched filtering.

9.6 Suppose the available peak power was unlimited. Define a CW (i.e. time-bandwidth-product = unity) signal which yields the same target detectability and range resolution as a LFM pulse of amplitude A, duration T, and bandwidth B. How do the ambiguity diagrams for the LFM and CW pulses differ?

9.7 Verify that (9.23) is correct. The addition and subtraction of αt in the exponent is necessary.

10

GENERALIZED COHERENT AND NONCOHERENT DETECTION AND PROCESSING

In what follows, coherent and noncoherent detection and processing are treated using an arbitrary complex signal. This serves to demonstrate the generality of the results of Chapter 6, in which a particular form of $s(t)$ was used.

In particular, referring to Section 7.1, the complex input signal and noise $z_s(t)$ and $z_n(t)$ are

$$z_s(t) = \psi(t)e^{i(2\pi f_0 t+\theta)}$$
$$z_n(t) = n(t) - i\hat{n}(t) \tag{10.1}$$

where θ is unknown, $s(t) = \mathrm{Re}(z_s(t))$, $n(t)$ is a realization of mean-zero white Gaussian noise, $\hat{n}(t)$ is the Hilbert transform (HT) of $n(t)$ and

$$\int_{-\infty}^{\infty} |\psi(t)|^2\, dt = 2E$$
$$E[n(t_1)n(t_2)] = \frac{N_0}{2}\,\delta(t_1 - t_2) \tag{10.2}$$

The complex amplitude $\psi(t)$ is assumed to be band-limited with $B \leq 2f_0$, which as discussed in Section 7.1, permits the form of $z_s(t)$ in (10.1). Also

$$E[\hat{n}(t_1)\hat{n}(t_2)] = \frac{N_0}{2}\,\delta(t_2 - t_1) \tag{10.3}$$

This holds because if

$$N(f) = \int_{-\infty}^{\infty} n(t)e^{-i2\pi ft}\, dt \tag{10.4}$$

then

$$\int_{-\infty}^{\infty} \hat{n}(t)e^{-i2\pi f_0 t}\, dt = -i \operatorname{sgn}(f)N(f) \tag{10.5}$$

where $\operatorname{sgn} f = \pm 1$ for $f \gtrless 0$. Thus the HT operation is equivalent to passing $n(t)$ through a filter with transfer function $-i \operatorname{sgn} f$. But by (3.11) the power spectral density at the output of such a filter is $|-i \operatorname{sgn} f|^2 = 1$ multiplied by the input power spectral density, which therefore remains unchanged. On the other hand, using (4.34) and (10.3),

$$E[n(t_1)\hat{n}(t_2)] = \frac{1}{\pi}\int_{-\infty}^{\infty} \frac{E[n(t_1)n(x)]}{t_2 - x}\, dx = \frac{N_0}{2\pi}\frac{1}{t_2 - t_1} \tag{10.6}^{\dagger}$$

The complex impulse response of the incoherent filter matched in amplitude-only to $z_s(t)$ is

$$z_s^*(T-t) = \psi^*(T-t)e^{-i2\pi f_0(T-t)} \tag{10.7}$$

where the arbitrary signal duration T is included for causality.

10.1 NONCOHERENT DETECTION OF A SINGLE PULSE

The complex quadrature receiver is diagrammed in Fig. 10.1, for which the complex filter output $z_y(t)$ is

$$\begin{aligned} z_y(t) &= e^{-i[2\pi f_0(T-t)-\theta]}\int_{-\infty}^{\infty} \psi(\tau)\psi^*(T-t+\tau)\, d\tau \\ &\quad + e^{-i2\pi f_0(T-t)}\int_{-\infty}^{\infty} z_n(\tau)\psi^*(T-t+\tau)e^{-i2\pi f_0\tau}\, d\tau \end{aligned} \tag{10.8}$$

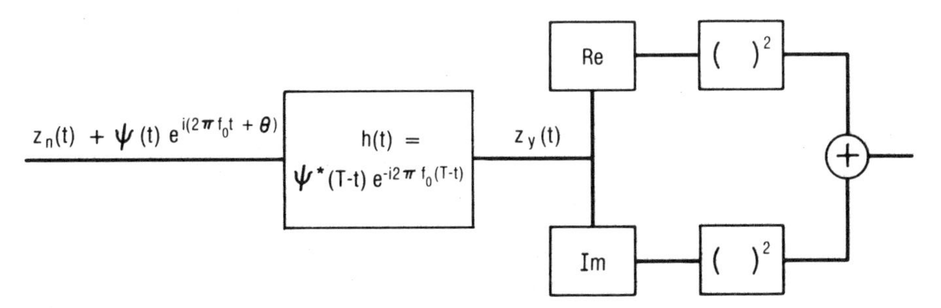

Figure 10.1 Generalized quadrature receiver for complex signals.

† The reader might wonder about the behavior of (10.6) for $t_1 = t_2$. This can be determined by considering white noise band limited to $|f| < B$ and letting $B \to \infty$. It can be shown that $E(n(t)\hat{n}(t+\tau)) = N_0(\sin \pi B\tau)^2/\pi\tau$ which vanishes at $\tau = 0$ for any value of B.

and the counterparts of the in-phase and quadrature components of the real quadrature receiver output at the time of signal correlation $t = T$ are respectively

$$\text{Re } z_y(T) = X = 2E\cos\theta + \text{Re}\int_{-\infty}^{\infty} z_n(\tau)\psi^*(\tau)e^{-i2\pi f_0\tau}\, d\tau = 2E\cos\theta + U$$

$$\text{Im } z_y(T) = Y = 2E\sin\theta + \text{Im}\int_{-\infty}^{\infty} z_n(\tau)\psi^*(\tau)e^{-i2\pi f_0\tau}\, d\tau = 2E\sin\theta + V \tag{10.9}$$

The random variables U and V are mean-zero, Gaussian and independent (see Exercise 1). The receiver output

$$z = (2E\cos\theta + U)^2 + (2E\sin\theta + V)^2 \tag{10.10}$$

therefore has a noncentral chi-square distribution with two degrees of freedom, which as is discussed in Section 6.4.2 takes the simple form

$$\frac{1}{2\sigma^2}\exp\left[-\frac{(z+\beta)}{2\sigma^2}\right] I_0\left[\frac{(z\beta)^{1/2}}{\sigma^2}\right] \tag{10.11}$$

where $\beta = 4E^2$ and $\sigma^2 = \text{Var}(U) = \text{Var}(V)$ which we now calculate.

The variance σ^2 is

$$\begin{aligned}\sigma^2 &= E\left[\frac{1}{2}\int_{-\infty}^{\infty} z_n(\tau)\psi^*(\tau)e^{-i2\pi f_0\tau}\, d\tau + \frac{1}{2}\int_{-\infty}^{\infty} z_n^*(\tau)\psi(\tau)e^{i2\pi f_0\tau}\, d\tau\right]^2 \\ &= \frac{1}{4}\int_{-\infty}^{\infty}\int_{-\infty}^{\infty} E[z_n(\tau_1)z_n(\tau_2)]\psi^*(\tau_1)\psi^*(\tau_2)e^{-i2\pi f_0(\tau_1+\tau_2)}\, d\tau_1\, d\tau_2 \\ &\quad + \text{complex conjugate} \\ &\quad + \frac{1}{4}\int_{-\infty}^{\infty}\int_{-\infty}^{\infty} E[z_n(\tau_1)z_n^*(\tau_2)]\psi^*(\tau_1)\psi(\tau_2)e^{-i2\pi f_0(\tau_1-\tau_2)}\, d\tau_1\, d\tau_2 \\ &\quad + \text{complex conjugate}\end{aligned} \tag{10.12}$$

The first two terms with $e^{\pm i2\pi f_0(\tau_1+\tau_2)}$ vanish (see Exercise 10.2). The third term is

$$\begin{aligned}&\frac{1}{4}\int_{-\infty}^{\infty}\int_{-\infty}^{\infty} E[(n(\tau_1) - i\hat{n}(\tau_1))(n(\tau_2) + i\hat{n}(\tau_2))] \\ &\quad \times \psi^*(\tau_1)\psi(\tau_2)e^{-i2\pi f_0(\tau_1-\tau_2)}\, d\tau_1\, d\tau_2\end{aligned} \tag{10.13}$$

Now, using (10.3) and (10.6)

$$E[n(\tau_1) - i\hat{n}(\tau_1)][n(\tau_2) + i\hat{n}(\tau_2)] = N_0\delta(\tau_1 - \tau_2) + \frac{iN_0}{\pi}\frac{1}{\tau_1 - \tau_2} \tag{10.14}$$

and (10.13) is therefore equal to

$$\frac{1}{4}\left[2EN_0 + iN_0 \int_{-\infty}^{\infty} \psi^*(\tau_1)e^{-i2\pi f_0 \tau_1}\, d\tau_1 \, \frac{1}{\pi}\int_{-\infty}^{\infty} \frac{\psi(\tau_2)e^{i2\pi f_0\tau_2}\, d\tau_2}{\tau_1 - \tau_2}\right] \quad (10.15)$$

But the integral on τ_2 is just the Hilbert transform of $\psi(\tau_2)e^{i2\pi f_0\tau_2}$ which, because $\psi(\tau_2)$ is band-limited as described above, is equal to $-i\psi(\tau_1)e^{i2\pi f_0\tau_1}$ (see Exercise 10.3). Therefore (10.15) is equal to

$$\tfrac{1}{4}(2EN_0 + 2EN_0) \quad (10.16)$$

which is real, and (10.12) yields

$$\sigma^2 = 2EN_0 \quad (10.17)$$

Hence, referring to Section 6.4.2

$$P_{\text{fa}} = \int_{\eta^2}^{\infty} \exp\left(-\frac{z}{4EN_0}\right)\frac{dz}{4EN_0} = \exp\left(-\frac{\eta^2}{4EN_0}\right) \quad (10.18)$$

and

$$P_{\text{d}} = \int_{\eta^2}^{\infty} \exp\left(-\frac{z+4E^2}{4N_0E}\right) I_0\left(\frac{z^{1/2}}{N_0}\right)\frac{dz}{4N_0E}$$
$$= \int_{(-2\ln P_{\text{fa}})^{1/2}}^{\infty} x \exp\left[-\frac{(x^2+\alpha^2)}{2}\right] I_0(x\alpha)\, dx \quad (10.19)$$

where $\alpha^2 = 2E/N_0$, which is identical to the results of Section 6.2 in which envelope rather than square-law detection was used. This further demonstrates the equivalence of monotonic detection schemes in the single-pulse case.

10.2 COHERENT AND NONCOHERENT INTEGRATION

With noncoherent integration the output of the square-law detector after integration is

$$S_M = \sum_{i=1}^{M} (2E\cos\theta_i + U_i)^2 + (2E\sin\theta_i + V_i)^2 = \sum_{i=1}^{M} R_i$$

where

$$U_i = \text{Re}\int_{-\infty}^{\infty} z_{n_i}(\tau)\psi^*(\tau)e^{-i2\pi f_0\tau}\, d\tau$$
$$V_i = \text{Im}\int_{-\infty}^{\infty} z_{n_i}(\tau)\psi^*(\tau)e^{-i2\pi f_0\tau}\, d\tau \quad (10.20)$$

and θ_i is allowed to vary from pulse to pulse.

Again, we apply the central limit theorem to the random variable

$$R_i = (2E\cos\theta_i + U_i)^2 + (2E\sin\theta_i + V_i)^2 \tag{10.21}$$

whose probability density approaches a Gaussian as M becomes large, with mean μ_M and variance σ_M^2 given by

$$\begin{aligned}\mu_M &= ME(R_i) = M(4E^2 + 4EN_0) = 4MEN_0\left(1 + \frac{E}{N_0}\right)\\ \sigma_M^2 &= M[E(R_i^2) - (E(R_i))^2]\end{aligned} \tag{10.22}$$

As before in Section 6.3.1, using (6.40)

$$\begin{aligned}E(U_i^2V_i^2) &= E(U_iU_iV_iV_i) = E(U_i^2)E(V_i^2) + 2(E(U_iV_i))^2\\ &= E(U_i^2)E(V_i^2) = 4E^2N_0^2\end{aligned} \tag{10.23}$$

Also by (6.40)

$$E(U_i^4) = 3E(U_i^2) = 12N_0^2E^2 = E(V_i^4) \tag{10.24}$$

Evaluation of σ_M^2, P_{fa} and P_{d} proceeds exactly as in Section 6.4.2 and

$$\sigma_M^2 = M16E^2N_0^2\left(1 + \frac{2E}{N_0}\right) \tag{10.25}$$

and

$$P_{\text{fa}} = \int_\eta^\infty \exp\left[-\frac{(x-\mu)^2}{2\sigma^2}\right]\frac{dx}{\sqrt{2\pi\sigma^2}} = \frac{1}{2}\,[1 - \text{erf}(\gamma)] \tag{10.26}$$

where μ and σ are obtained from (10.22) and (10.25) by setting $E/N_0 = 0$. In this case the threshold η is

$$\eta = 4MN_0E\left(1 + \gamma\sqrt{\frac{2}{M}}\right) \tag{10.27}$$

and, as before

$$P_{\text{d}} = \frac{1}{2}\left[1 + \text{erf}\left(\frac{\sqrt{\frac{M}{2}}\,\frac{E}{N_0} - \gamma}{\left(1 + \frac{2E}{N_0}\right)^{1/2}}\right)\right] \tag{10.28}$$

For coherent integration the output of the square-law detector is

$$\left(2ME\cos\theta + \sum_{i=1}^{M} U_i\right)^2 + \left(2ME\sin\theta + \sum_{i=1}^{M} V_i\right)^2$$
$$= (B_1 + x_1)^2 + (B_2 + x_2)^2 \tag{10.29}$$

which, again, has a chi-square distribution with two degrees of freedom with $B_1 = 2ME\cos\theta$, $B_2 = 2ME\sin\theta$, $x_1 = \Sigma_{i=1}^{M} U_i$, $x_2 = \Sigma_{i=1}^{M} V_i$ and it is assumed that θ remains constant over the integration time.

In this case

$$P_{\mathrm{d}} = \int_{\eta^2}^{\infty} \exp\left(-\frac{z+\beta}{2\sigma^2}\right) I_0\left[\frac{(z\beta)^{1/2}}{\sigma^2}\right] \frac{dz}{2\sigma^2} \tag{10.30}$$

where

$$\beta = 4M^2E^2\,, \quad \sigma^2 = 2MEN_0\,, \quad \eta = (-2\sigma^2 \ln P_{\mathrm{fa}})^{1/2}$$

which yields

$$P_{\mathrm{d}} = \int_{(-2\ln P_{\mathrm{fa}})^{1/2}}^{\infty} x \exp\left(-\frac{x^2+\alpha^2}{2}\right) I_0(x\alpha)\, dx \tag{10.31}$$

where $\alpha^2 = 2ME/N_0$ which, again, is identical to the result in Section 6.4.2.

EXERCISES FOR CHAPTER 10

10.1 Show by using (10.1) that U and V in (10.10) are independent.

10.2 Show by using (10.1) that

$$\frac{1}{4}\int_{-\infty}^{\infty}\int_{-\infty}^{\infty} E[z_n(\tau_1)z_n(\tau_2)]\psi^*(\tau_1)\psi^*(\tau_2)e^{-i2\pi f_0(\tau_1+\tau_2)}\, d\tau = 0$$

10.3 Show that

$$\frac{1}{\pi}\int_{-\infty}^{\infty} \frac{\psi(\tau_1)e^{i2\pi f_0\tau_2}\, d\tau_2}{\tau_1 - \tau_2} = -i\psi(\tau_2)e^{i2\pi f_0\tau_1}$$

by using the fact that $\psi(\tau_2)$ is a band-limited function, with $B < 2f_0$.

11

SYSTEMS CONSIDERATIONS

11.1 BEAMPATTERNS AND GAIN OF ANTENNAS AND ARRAYS

Consider an idealized one-dimensional reflector-type antenna with aperture-width D as diagrammed in Figure 11.1. In the vicinity of the reflector (the near field) the radiation is collimated and at large distances (the far field) it becomes spread out due to diffraction with, as will be shown, most of the energy being confined to an angular range λ/D, where λ is the wavelength. The far field radiation pattern is of interest here, and the distance R that may be taken as the boundary for the beginning of the far field can be determined from the accompanying geometrical diagram due to Lord Rayleigh. It is seen that for distances greater than D^2/λ the width of the radiation field corresponding to the far-field angular width λ/D exceeds that of the collimated beam. We therefore say that for $R > D^2/\lambda$ we are in the far field of the antenna.

In order to calculate the far field pattern each incremental element dx over the aperture of the reflector is treated as a source of radiation† of the form $A(x)\exp[i(2\pi f_0 t + \phi(x)]\,dx$ where $f_0 = c/\lambda$ is the carrier frequency and c the propagation speed. The field $F(P)$ at some point P at range R will contain the contributions from each element over the aperture and be of the form

$$F(P) = K\int_{-D/2}^{D/2} A(x)\,\frac{\exp\left[i\left(2\pi f_0\left(t - \frac{R}{c}\right) + \phi(x)\right)\right]}{R}\,dx \qquad (11.1)$$

† Huygeni Principle.

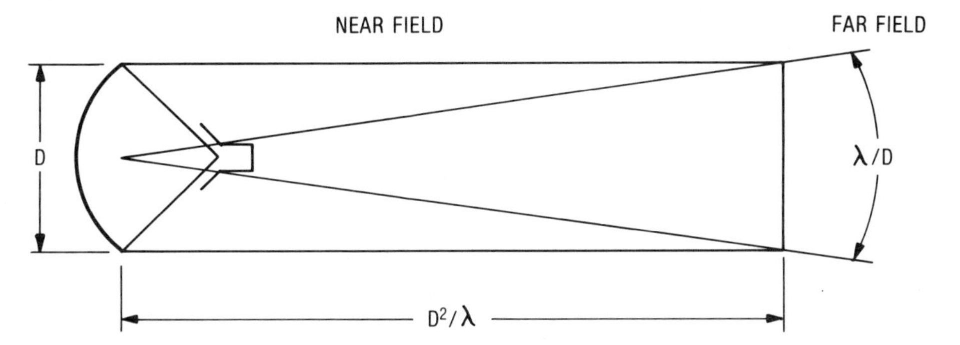

Figure 11.1 Approximate near field and far field radiation patterns.

where the factor $1/R$ accounts for attenuation of the field amplitude with distance and K is a complex constant.

From Figure 11.2

$$R = (R_0^2 - 2xx' + x^2)^{1/2} = R_0\left(1 - \frac{2xx'}{R_0^2} + \frac{x^2}{R_0^2}\right)^{1/2}$$

$$\sim R_0 + \frac{x^2}{2R_0} - \frac{xx'}{R_0}\left(1 + \frac{xx'}{2R_0^2}\right) + \cdots \tag{11.2}$$

and the phase term involving R in the exponent of (11.1) is therefore

$$2\pi \frac{R_0}{\lambda} + \frac{x^2}{2\lambda R_0} - \frac{xx'}{\lambda R_0}\left(1 + \frac{xx'}{2R_0^2}\right) + \cdots$$

Now in practice

$$\frac{x^2}{2\lambda R_0} \leq \frac{D^2}{2\lambda R_0}, \quad \frac{x'}{R_0} \approx \sin\theta \leq 1 \tag{11.3}$$

hence, for $R_0 \gg D^2/\lambda$ the terms $x^2/2\lambda R_0$ and $xx'/2R_0^2$ in the exponent can be ignored. Also to a good approximation the attenuation factor $1/R$ can be

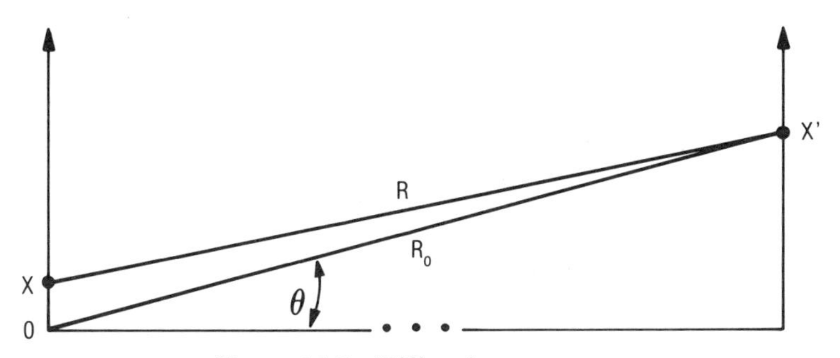

Figure 11.2 Diffraction geometry.

replaced by $1/R_0$, which yields the Fraunhofer diffraction integral

$$F(\theta)=\frac{\exp\left[i2\pi\left(f_0t+\frac{R_0}{\lambda}\right)\right]}{R_0}K\int_{-D/2}^{D/2}A(x)\exp\left[-i\left(2\pi\frac{x\sin\theta}{\lambda}+\phi(x)\right)\right]dx \tag{11.4}$$

which is the Fourier transform, in terms of the variable $\sin\theta/\lambda$, of the aperture amplitude and phase distribution $A(x)e^{i\phi(x)}$.

Consider a uniform amplitude and phase distribution, $A(x)e^{i\theta(x)}=1$. The magnitude of the response is of interest here which in this case is, ignoring K/R_0

$$|f(\theta)|=\left|\frac{\sin\pi D(\sin\theta/\lambda)}{\pi(\sin\theta/\lambda)}\right| \tag{11.5}$$

The extension to a two-dimensional rectangular reflector with dimensions (D_x, D_y), and angles $\sin\theta_x=x'/R_0$, $\sin\theta_y=y'/R_0$ is immediate and given by

$$|f(\theta_x,\theta_y)|=\left|\frac{\sin\pi D_x(\sin\theta_x/\lambda)}{\pi(\sin\theta_x/\lambda)}\right|\left|\frac{\sin\pi D_y(\sin\theta_y/\lambda)}{\pi(\sin\theta_y/\lambda)}\right| \tag{11.6}$$

and a circular aperture of diameter D with uniform illumination yields the Airy pattern

$$f(\theta)=\left[\frac{2J_1(\pi D\sin\theta)/\lambda)}{\pi D(\sin\theta/\lambda)}\right]$$

where J_1 is the ordinary Bessel function of first order.

In exact parallel with the discussion in Section 4.3 dealing with the spectrum of a rectangular pulse the half-power beamwidth $\Delta\sin\theta$ of the uniformly illuminated rectangular aperture of width D is from the foregoing results

$$(\Delta\sin\theta)\approx\Delta\theta=\frac{\lambda}{D} \tag{11.7}$$

since in practice $\lambda\ll D$. For a circular aperture the beamwidth is slightly larger, given by

$$\Delta\theta\sim\frac{4}{\pi}\frac{\lambda}{D} \tag{11.8}$$

Array antennas differ from continuous-aperture antennas in that they are made up of a number of discrete antenna elements which are operated collectively. The discrete elements can be anything from simple dipoles to large parabolic reflectors. For arrays there are results similar to those for continuous antennas, but there are also important differences. Consider an array of isotropic radiators as illustrated in Figure 11.3. Let each radiator be excited by a signal $e^{i2\pi ft}$ and consider the field at a point P at a distance R in the far field, with angle of incidence to the array θ. As $R \to \infty$ the rays drawn from each radiator to P are all parallel and the relative phase shifts in the contributions from elements at $\pm nd$ on either side of the element at the center are $\pm 2\pi nd \sin\theta/\lambda$. The field amplitude at P is therefore, assuming for convenience that N is odd,

$$F(\theta) = \frac{K}{R} \sum_{n=-(N-1)/2}^{(N-1)/2} \exp\left[i2\pi\left(f_0 t + \frac{nd \sin\theta}{\lambda}\right)\right]$$

where K is a complex constant and the magnitude of the response is, again ignoring K/R

$$|f(\theta)| = \left| \frac{\sin \pi N(d \sin\theta/\lambda)}{\sin \pi(d \sin\theta/\lambda)} \right| \tag{11.9}$$

where we have used

$$\sum_{n=(N-1/2)}^{(N-1/2)} e^{inz} = \frac{\sin \pi Nz}{\sin \pi z} \tag{11.10}$$

The half-power beamwidth is therefore, as before:

$$\Delta \sin\theta \sim \Delta\theta = \frac{\lambda}{Nd} \sim \frac{\lambda}{L} \tag{11.11}$$

where $L = (N-1)$, d is the length of the array, which is essentially identical to (11.7). The pattern of $f(\theta)$ has a principal maximum at $\sin\theta = 0$, but, in addition, there are also maxima for $\sin\theta = \lambda/d,\ 2\lambda/d,\ 3\lambda/d, \ldots,$ which continuous radiation sources do not have. These secondary and generally unwanted maxima are referred to as grating lobes, and are mathematically equivalent to the Doppler ambiguities considered in Section 8.1, as will be discussed.

On reception, for a radiation source in the far field located broadside to the array, that is $\theta = 0$, the phase relationships between the signals at the outputs of the elements will be identical to those in the preceding discussions for transmission. Also, for $\theta \neq 0$ the phase difference between adjacent elements will, as before, be $2\pi d \sin\theta/\lambda$. Thus arrays, and in fact all antennas, are reciprocal, in the sense that the beampatterns are the same for transmission or reception. Off broadside, transmission or reception of energy for some angular location $\theta_0 \neq 0$ is accomplished by implementing a phase shift of $\pm n(d/\lambda) \sin\theta_0$ radians at the input (output) of each element located at a distance $\pm nd$ from the center of the array, in which case the

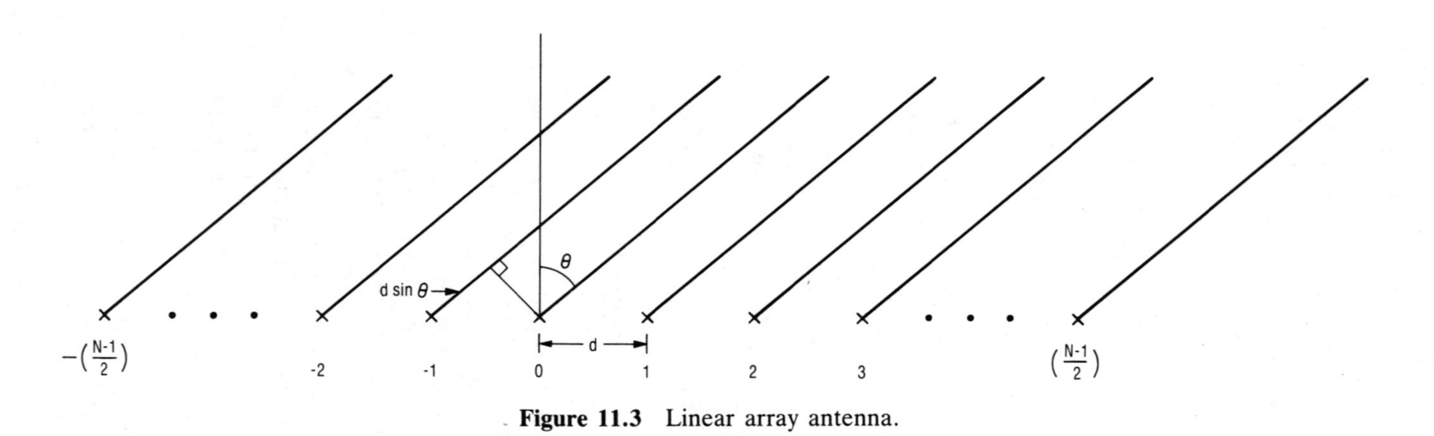

Figure 11.3 Linear array antenna.

transmission or reception beampattern is

$$|f(\theta)| = \frac{\sin(\pi N d/\lambda)(\sin\theta - \sin\theta_0)}{\sin(\pi d/\lambda)(\sin\theta - \sin\theta_0)} \tag{11.12}$$

In this discussion the central element has been used as the phase-reference point, but of course the phase can be referred to any point on the array with no change in the magnitude of the beam pattern. In this way linear and planar arrays are steered electronically. For reflector-type antennas steering is of course accomplished by physically pointing the antenna in the appropriate direction.

Now returning to the subject of grating lobes, if an array has grating lobes in the visible half space, $-90° \le \theta \le 90°$, there is essentially no way of determining whether a source of radiation in the far field is at an angular location corresponding to the principal maximum or at an angular location $\theta = \sin^{-1}[n(\lambda/d)]$, $n = \pm 1, \pm 2, \ldots$ on either side of it. Also, transmitted energy becomes distributed over the grating lobes rather than being confined to a single desired direction. However, grating lobes can be effectively eliminated as follows. We note that if the lobes can be separated by π radians, which corresponds to forward and backward radiation, then there is only a single such lobe, which only just begins to enter the visible half space (at $\mp 90°$) when the principal maximum is steered to end-fire, at $\pm 90°$. This requires that $\sin \pi[(d/\lambda)\sin \pi/2] = 1$, and $d = \lambda/2$. Thus, if $d < \lambda/2$ grating lobes are effectively eliminated. Note that for $d = \lambda/2$, the array beamwidth is

$$\frac{\lambda}{(N-1)d} = \frac{2}{N-1} \text{ radians} \approx \frac{115}{N} \text{ degrees} \tag{11.13}$$

In any sensing system it is desirable to be able to concentrate the transmitted power as much as possible towards a specific point in space, and also to enhance the sensitivity of the sensor, on reception, for signals from a particular direction. This increases the range at which a target at a particular point can be detected—recall the factor $1/R_0$ in (11.4)—and enables the angular location of a scatterer to be more accurately identified. By the reciprocity property of antennas these two capabilities are equivalent and are minimally satisfied by an isotropic radiator, which has a uniform angular response over 4π stearadians for transmission and reception. On the other hand, for a radiator with a rectangular aperture of dimensions D_x and D_y it has been shown that transmitted radiation is confined for the most part to a solid angle of

$$\frac{\lambda^2}{D_x D_y} \sim \frac{\lambda^2}{A} \text{ stearadians} \tag{11.14}$$

where A is the area of the aperture. We therefore define the gain of an antenna relative to an isotropic radiator as

$$G = \frac{4\pi}{\text{solid-angle beanwidth}} = \frac{4\pi A}{\lambda^2} \tag{11.15}$$

To repeat, G is a measure of capability for selecting a given localized region in space both for transmission and reception of energy. Gain increases directly with the area of the aperture, and the region becomes more localized and the antenna angular response more selective with increase in the gain. Note that the maximum value of (11.6) is $D_x D_y = A$ which occurs at $\theta_x = \theta_y = 0$. Thus the gain of the antenna is proportional to the maximum value of the beampattern. Equation (11.15) holds equally for two-dimensional arrays as well as reflectors of all kinds.

Arrays, when used in radar applications are usually two-dimensional, and the extension of (11.9) and (11.12) to the rectangular case is immediate, and exactly analogous to (11.6). In sonar however it is more customary at present to employ linear arrays, and (11.9) and (11.12) apply directly. The question of array gain in sonar is also handled somewhat differently. A linear array of N elements receiving a plane wave representing scattering from a point target in the far field, yields a signal amplitude N times that of a single element, because the signal amplitude is coherent from element to element. The signal gain in radar would therefore be proportional to $20 \log N$. In sonar however one often deals with the gain in signal-to-noise ratio rather than just signal power alone. In this case, assuming spatially white ambient noise—that is, independent from element to element—the noise power at the output is $10 \log N$. We therefore have the array gain AG for the received signal

$$\text{AG} = 20 \log N - 10 \log N = 10 \log N$$

which is the improvement in *signal-to-noise ratio* over that which would be achieved by a single array element.

11.2 THE RADAR AND SONAR EQUATIONS

Let a target at range R from a measurement point be illuminated by a plane wave E_{inc}. If the target scatters energy uniformly over 4π stearadians and E_{sc} is the scattered wave at the location of the target, the received power P is

$$P = \frac{|E_{\text{sc}}|^2}{4\pi R^2} \tag{11.16}$$

The scattered field at the target E_{sc} can be written in terms of the incident field as $|E_{\text{sc}}|^2 = \sigma |E_{\text{inc}}|^2$ where the constant of proportionality σ is the scattering cross section, which has dimensions of area and in terms of the received power is given by

$$\sigma = 4\pi \frac{P}{|E_{\text{inc}}|^2} R^2 \tag{11.17}$$

In most cases of interest targets do not scatter uniformly over 4π stearadians and σ is therefore a function of the angle between the incident field vector and the vector pointing from the target towards the receiver. In radar the factor of 4π is nevertheless retained, and the radar cross section is therefore a measure of the effective area of an equivalent isotropic scatterer capable of scattering 4π times the amount of power per unit solid angle that is actually scattered towards the receiver. In general the angular directions to the receiver and to the transmitter may not coincide, in which case we speak of bistatic scattering and bistatic-scattering cross sections $\sigma(|\mathbf{k}_{\text{I}} - \mathbf{k}_{\text{S}}|)$, where $|\mathbf{k}_{\text{I}} - \mathbf{k}_{\text{S}}|$ represents the angle between the incident and scattered wave vectors. Here however we shall deal only with backscatter—for which the transmitter and receiver coincide—and denote σ, with no argument, as the backscatter cross section.

In sonar the equivalent quantity is known as the Target Strength, which is related to σ but defined somewhat differently. For an incident wave e^{ikx}, where $k = 2\pi/\lambda$, the amplitude W of the scattered wave in the far field at a distance R can be written as

$$W = \frac{f(|\mathbf{k}_{\text{I}} - \mathbf{k}_{\text{S}}|)e^{ikx}}{R} \tag{11.18}$$

where $f(|\mathbf{k}_{\text{I}} - \mathbf{k}_{\text{S}}|)$ describes the scattering properties of the target as a function of the angle between the incident and scattered waves. The function $f(|\mathbf{k}_{\text{I}} - \mathbf{k}_{\text{S}}|)$ has dimensions of length and is related to σ by

$$\sigma = 4\pi |f|^2 \tag{11.19}$$

In sonar the dimensionless target strength TS is defined in terms of W in (11.18), with R set equal to 1 m, as

$$\text{TS} = 10 \log |W|^2 = 10 \log |f|^2 \text{ dB} \tag{11.20}$$

Thus TS is a measure of the scattered power referred to 1 m from the scattering point in the direction of interest.

In transmission, in a radar system a voltage is converted into an electromagnetic wave which is radiated by the antenna. In sonar the electrical signal is converted via an electroacoustic transducer into an acoustic signal which is a mechanical vibration. Let us consider the former case first. Let P_{T} denote the peak transmitted power; for a rectangular pulse of amplitude v this would be v^2. If the antenna gain is G, the power per stearadian in the direction of the maximum of the beam pattern is

$$\frac{P_T G}{4\pi} \tag{11.21}$$

The solid angle subtended at the transmitter/receiver by a target of radar cross section (RCS) σ at range R is σ/R^2 and the power intercepted by the target is therefore

$$\frac{P_T G \sigma}{4\pi R^2} \tag{11.22}$$

Since RCS is defined in terms of an equivalent isotropic radiator, the scattered power per unit area at the receiver is

$$\frac{P_T G \sigma}{4\pi R^2} \frac{1}{4\pi R^2} \tag{11.23}$$

The power collected by the antenna is the area A multiplied by (11.23), and using (11.15) the received power P can be written as

$$P = \frac{P_T \lambda^2 G^2 \sigma}{(4\pi)^3 R^4} \tag{11.24}$$

which is sometimes more convenient to express in terms of the area of the antenna as

$$P = \frac{P_T A^2 \sigma}{4\pi \lambda^2 R^4} \tag{11.25}$$

Note that as G increases, and the beamwidth becomes narrower, the collecting area increases.

The most important features of the radar equation, (11.24) or (11.25), are: the $1/R^4$ dependence which incorporates the two-way $1/R^2$ attenuation that the signal undergoes in propagating up and back; the dependence on the area of the aperture; and the fact that, for fixed A, P increases with increase in frequency $f = c/\lambda$. In (11.25) A should be taken as the effective aperture which is the actual geometrical aperture multiplied by an efficiency factor which may range anywhere between 0.3 and 0.9.

For a rectangular pulse of magnitude v—either a CW pulse with $BT = 1$ or a large time-bandwidth rectangular pulse with $BT \gg 1$ such as those discussed in Chapter 9—the transmitted energy E_T is $P_T T = v^2 T/2$. Therefore, by multiplying both sides of (11.25) by T/N_0, where $N_0/2$ is the two sided noise power spectral density, the ratio of received signal energy to noise spectral density, which is the signal-to-noise ratio at the output of an incoherent matched filter, is

$$\frac{E}{N_0} = \frac{E_T A^2 \sigma}{N_0 4\pi\lambda^2 R^4} \tag{11.26}$$

In (11.26) the value of N_0 should represent the noise due to the effective antenna temperature plus the overall system noise characteristics, which includes the front-end amplifier and noise generated in lossy elements, which can be calculated using the results of Section 3.7. From (3.59) it is clear that in a properly designed system the overall system noise characteristics are established by the noise generated in the front end.

Derivation of a corresponding equation for sonar is much less straightforward owing to the more complex propagation environment in the ocean. In radar spherical spreading of the electromagnetic waves can be assumed, but in sonar this holds only for relatively short distances from the transmitter, before reflections from the sea surface and the bottom come into play. Also, the depth dependence of the sound speed is generally such that a duct or channel is created within which a sonar signal may be confined over long distances by refraction. Ducting can also occur in radar, but is the exception rather than the rule. Thus, a sonar signal propagation path can involve bounces between the surface and the bottom, channeling due to refraction, and also hybrid paths consisting of a top and/or a bottom reflection plus refraction [6, 7]. As a result, propagation can be similar to that in a waveguide with spreading of the signal becoming cylindrical rather than spherical [35], and amplitude attenuation by a factor of $1/\sqrt{R}$ rather than $1/R$, which can result in very long range propagation.

Let us now consider a monostatic sonar system employing a linear array of N elements. A given amount of electrical power input to each element will produce a pressure wave of some amplitude and some duration T in the water. It is customary to work with logarithmic quantities (dB) with a reference pressure of 1 microPascal (μPa; 1 μPa $= 10^{-5}$ dyn/cm^2). Denote the pressure wave in dB produced by each element as SL (source level). Since there are N elements, there will be a signal power gain of $20 \log N$ in the direction of interest over a single isotropic radiator. In propagating to the target there will be a transmission loss TL and in scattering there will be a gain by a factor of the target strength, TS. In the return path there is an additional loss of TL and at the array a gain in SNR of $10 \log N$ which, as discussed above, assumes spatially white ambient noise. The signal-to-noise ratio (SNR) of the output of the array, in the sense given by (11.26), is therefore given by

$$\text{SNR} = \text{SL} + 10 \log T + 30 \log N - 2\text{TL} + \text{TS} - 10 \log N_0 \tag{11.27}$$

where the power spectral density N_0 here represents ambient noise and noise generated in the receivers.

In using (11.27) it is necessary to evaluate TL, which is dependent on the particular situation of interest and will generally require use of computer codes, ray tracing routines, etc. Also, (11.27) does not include effects of reverberation, which in sonar is the counterpart of radar clutter. Reverberation however is usually much more of a serious problem than clutter for a radar, owing to the presence of the surface and bottom of the ocean as well as the presence of shipping, marine life, etc., which is exacerbated by the long-range acoustic propagation.

11.3 THE SEARCH PROBLEM

The results of the preceding sections express operational capabilities of sensors in terms of the peak power that can be generated. The average power however is of equal importance, particularly in multipulse functions such as signal integration, which has already been mentioned, and in search, which we now consider for radar. From (11.26) the range at which a radar is capable of realizing a certain value of (E/N_0) for a single pulse is

$$R = \left[\frac{A^2(E_T/N_0)\sigma}{4\pi\lambda^2(E/N_0)} \right]^{1/4} \tag{11.28}$$

Now suppose the sensor is to search in some time T_S over a volume defined by a solid angle Ω and a range R. The volume search rate SR can be defined as

$$\mathrm{SR} = \frac{\Omega R^3}{3T_S} \tag{11.29}$$

We can express the search angle approximately in numbers of antenna beamwidths as

$$\Omega \approx N \frac{\lambda^2}{A} \tag{11.30}$$

where λ^2/A is the solid-angle antenna beamwidth of the sensor and N is an integer. The search over the volume can be accomplished by transmitting a periodic waveform consisting of a pulse of duration T every T_p seconds, pointing the sensor during the time interval T_S to N different discrete beam positions which cover Ω, and dwelling for a time T_p at each beam position in order to allow the scattering from a possible target at the maximum range R to be received. We require[†]

$$T_p = \frac{T_S}{N} \tag{11.31}$$

[†] The constraint of (8.1) must of course also be satisfied.

Now using (11.28), (11.29) and (11.30)

$$\mathrm{SR} = \frac{1}{3R}\frac{\Omega R^4}{T_{\mathrm{S}}} = \frac{1}{3R}\frac{N\lambda^2}{A}\frac{A^2(E_{\mathrm{T}}/N_0)\sigma}{4\pi\lambda^2(E/N_0)NT_{\mathrm{p}}}$$

$$= \frac{\sigma}{12\pi RN_0}\frac{(P_{\mathrm{T}}T/T_{\mathrm{p}})A}{(E/N_0)} \tag{11.32}$$

since $E_{\mathrm{T}} = P_{\mathrm{T}}T$. The ratio T/T_{p} of pulse duration to interpulse spacing is the duty factor and $P_{\mathrm{T}}T/T_{\mathrm{p}}$ is the average power P_{av}, and we can write

$$\mathrm{SR} = \frac{\sigma}{12\pi RN_0}\frac{P_{\mathrm{av}}A}{(E/N_0)} \tag{11.33}$$

Thus, search capability is dependent on the power-aperture product, $P_{\mathrm{av}}A$ of the sensor, where P_{av} is the average power capability. The dependence on the other parameters is as might be expected. Although the search scenario chosen is not the only possibility, the results remain substantially unchanged. The noise power spectral density N_0 has been written separately from E/N_0 because the latter quantity may be specified as a separate numerical parameter dependent on required values of P_{d} and P_{fa}.

11.4 SPECIFICATION OF THE FALSE ALARM PROBABILITY, P_{fa}

The results of Chapters 5 and 6, in which sensor operation and capability is analyzed and evaluated in terms of detection and false alarm probabilities, all refer to single observations. That is, they deal with the probability that at any given instant the detection threshold is or is not exceeded, given that a target is or is not present. The threshold is dependent on the false alarm probability P_{fa} and we now deal with the question of how P_{fa} might be specified. As a general rule, issues related to detection and false alarm probabilities are most important during search; that is, prior to positive identification and acquisition of a target. How this is accomplished varies with the application. Typically, an exceedance of a threshold associated with illumination of a point in space is followed by re-interrogation of that point—or more precisely a small volume around the point in order to allow for target motion. A voting procedure then might be implemented in which, say, at least m exceedances out of n successive trials are required in order for the observation to be taken as sufficient evidence of a positive sighting. Sometimes the detection threshold is changed during the voting process in order to further reduce the possibility of a false acquisition. Clearly there is no general rule for specifying how a particular value of P_{fa} might be arrived at. However, since this issue is often most important during search, let us consider the following example which gives a flavor of how this might be done.

Let there be the requirement that a certain volume in space is to be searched in a time T_S and that during some fraction η of each search interval T_S—that is during a time ηT_S—the probability that there shall be not more than one false alarm is p; the value of p might typically be 0.999. This may be stated mathematically as:

$$P[\text{no false alarms in } \eta T_S \text{ seconds}] + P[\text{1 false alarm in } \eta T_S \text{ seconds}] = p \tag{11.34}$$

Thus, the problem of specifying p has been substituted for the problem of specifying P_{fa}, but the number of false alarms per search interval has more direct operational and physical meaning than does P_{fa}. Now, if the bandwidth of the signal is B it can be assumed that all observations are made at the output of a baseband filter of bandwidth B, and during a time T_S there are therefore $2BT_S$ opportunities for a false alarm to occur. That is, in Chapter 5 it was shown that noise observations at the output of a low-pass filter of bandwidth B that are separated in time by $1/2B$ seconds are independent and therefore $2B$ independent noise observations per second are possible. Therefore during each search interval T_S there are $2BT_S$ possibilities for an exceedance of the threshold. This result applies generally, under all conditions. Hence, the average number of false alarms during the interval ηT_S is $2\eta T_S BP_{fa}$.

Now referring to Section 2.2, it has been shown that the Poisson distribution can be applied when λ/n is small, where λ is the expected value of the number of events which occur during n Bernoulli trials, and λ/n is the probability of the occurrence of the event. In this case the event is the occurrence of a false alarm and relevant quantity is $2\eta T_S BP_{fa}/2\eta T_S B = P_{fa}$. Let us assume that $P_{fa} \leq 10^{-5}$. Poisson statistics can therefore be applied, and referring to (2.11) and (11.34), and denoting $x = 2\eta T_S BP_{fa}$ we have

$$e^{-x}(1 + x) = p \tag{11.35}$$

Typical values of these parameters are $T_S = 10^{-4}$ s, $B = 100$ MHz, and if $P_{fa} \leq 10^{-5}$ then $x \leq 1$. If for any reason this does not hold (11.35) can easily be solved for x by trial and error. In this example however we can write

$$e^{-x}(1 + x) \sim (1 - x)(1 + x) = 1 - x^2 = p \tag{11.36}$$

and

$$P_{fa} = \frac{(1 - p)^{1/2}}{2\eta BT_S} \tag{11.37}$$

For $p = 0.999$, $\eta = 1$ and the foregoing value of $BT_S = 10^4$, we find that

$$P_{fa} \approx 3 \times 10^{-6}$$

It is always important to check the self-consistency of assumptions which have been made, which of course is satisfied in this case since $P_{fa} < 10^{-5}$ was assumed.

APPENDIX

Table of Values of the Error Function

$$\mathrm{Erf}\,(x) = \frac{2}{\sqrt{\pi}} \int_0^x e^{-t^2} dt$$

x	0	1	2	3	4	5	6	7	8	9
0.00	0.00 000	00 113	00 226	00 339	00 451	00 564	00 677	00 790	00 903	01 016
0.01	0.01 128	01 241	01 354	01 467	01 580	01 692	01 805	01 918	02 031	02 144
0.02	0.02 256	02 369	02 482	02 595	02 708	02 820	02 933	03 046	03 159	03 271
0.03	0.03 384	03 497	03 610	03 722	03 835	03 948	04 060	04 173	04 286	04 398
0.04	0.04 511	04 624	04 736	04 849	04 962	05 074	05 187	05 299	05 412	05 525
0.05	0.05 637	05 750	05 862	05 975	06 087	06 200	06 312	06 425	06 537	06 650
0.06	0.06 762	06 875	06 987	07 099	07 212	07 324	07 437	07 549	07 661	07 773
0.07	0.07 886	07 998	08 110	08 223	08 335	08 447	08 559	08 671	08 784	08 896
0.08	0.09 008	09 120	09 232	09 344	09 456	09 568	09 680	09 792	09 904	10 016
0.09	0.10 128	10 240	10 352	10 464	10 576	10 687	10 799	10 911	11 023	11 135
0.10	0.11 246	11 358	11 470	11 581	11 693	11 805	11 916	12 028	12 139	12 251
0.11	0.12 362	12 474	12 585	12 697	12 808	12 919	13 031	13 142	13 253	13 365
0.12	0.13 476	13 587	13 698	13 809	13 921	14 032	14 143	14 254	14 365	14 476
0.13	0.14 587	14 698	14 809	14 919	15 030	15 141	15 252	15 363	15 473	15 584
0.14	0.15 695	15 805	15 916	16 027	16 137	16 248	16 358	16 468	16 579	16 689
0.15	0.16 800	16 910	17 020	17 130	17 241	17 351	17 461	17 571	17 681	17 791
0.16	0.17 901	18 011	18 121	18 231	18 341	18 451	18 560	18 670	18 780	18 890
0.17	0.18 999	19 109	19 218	19 328	19 437	19 547	19 656	19 766	19 875	19 984
0.18	0.20 094	20 203	20 312	20 421	20 530	20 639	20 748	20 857	20 966	21 075
0.19	0.21 184	21 293	21 402	21 510	21 619	21 728	21 836	21 945	22 053	22 162
0.20	0.22 270	22 379	22 487	22 595	22 704	22 812	22 920	23 028	23 136	23 244
0.21	0.23 352	23 460	23 568	23 676	23 784	23 891	23 999	24 107	24 214	24 322
0.22	0.24 430	24 537	24 645	24 752	24 859	24 967	25 074	25 181	25 288	25 395
0.23	0.25 502	25 609	25 716	25 823	25 930	26 037	26 144	26 250	26 357	26 463
0.24	0.26 570	26 677	26 783	26 889	26 996	27 102	27 208	27 314	27 421	27 527
0.25	0.27 633	27 739	27 845	27 950	28 056	28 162	28 268	28 373	28 479	28 584
0.26	0.28 690	28 795	28 901	29 006	29 111	29 217	29 322	29 427	29 532	29 637
0.27	0.29 742	29 847	29 952	30 056	30 161	30 266	30 370	30 475	30 579	30 684
0.28	0.30 788	30 892	30 997	31 101	31 205	31 309	31 413	31 517	31 621	31 725
0.29	0.31 828	31 932	32 036	32 139	32 243	32 346	32 450	32 553	32 656	32 760
0.30	0.32 863	32 966	33 069	33 172	33 275	33 378	33 480	33 583	33 686	33 788
0.31	0.33 891	33 993	34 096	34 198	34 300	34 403	34 505	34 607	34 709	34 811
0.32	0.34 913	35 014	35 116	35 218	35 319	35 421	35 523	35 624	35 725	35 827
0.33	0.35 928	36 029	36 130	36 231	36 332	36 433	36 534	36 635	36 735	36 836
0.34	0.36 936	37 037	37 137	37 238	37 338	37 438	37 538	37 638	37 738	37 838
0.35	0.37 938	38 038	38 138	38 237	38 337	38 436	38 536	38 635	38 735	38 834
0.36	0.38 933	39 032	39 131	39 230	39 329	39 428	39 526	39 625	39 724	39 822
0.37	0.39 921	40 019	40 117	40 215	40 314	40 412	40 510	40 608	40 705	40 803
0.38	0.40 901	40 999	41 096	41 194	41 291	41 388	41 486	41 583	41 680	41 777
0.39	0.41 874	41 971	42 068	42 164	42 261	42 358	42 454	42 550	42 647	42 743
0.40	0.42 839	42 935	43 031	43 127	43 223	43 319	43 415	43 510	43 606	43 701
0.41	0.43 797	43 892	43 988	44 083	44 178	44 273	44 368	44 463	44 557	44 652
0.42	0.44 747	44 841	44 936	45 030	45 124	45 219	45 313	45 407	45 501	45 595
0.43	0.45 689	45 782	45 876	45 970	46 063	46 157	46 250	46 343	46 436	46 529
0.44	0.46 623	46 715	46 808	46 901	46 994	47 086	47 179	47 271	47 364	47 456
0.45	0.47 548	47 640	47 732	47 824	47 916	48 008	48 100	48 191	48 283	48 374
0.46	0.48 466	48 557	48 648	48 739	48 830	48 921	49 012	49 103	49 193	49 284
0.47	0.49 375	49 465	49 555	49 646	49 736	49 826	49 916	50 006	50 096	50 185
0.48	0.50 275	50 365	50 454	50 543	50 633	50 722	50 811	50 900	50 989	51 078
0.49	0.51 167	51 256	51 344	51 433	51 521	51 609	51 698	51 786	51 874	51 962

x	0	1	2	3	4	5	6	7	8	9
0.50	0.52 050	52 138	52 226	52 313	52 401	52 488	52 576	52 663	52 750	52 837
0.51	0.52 924	53 011	53 098	53 185	53 272	53 358	53 445	53 531	53 617	53 704
0.52	0.53 790	53 876	53 962	54 048	54 134	54 219	54 305	54 390	54 476	54 561
0.53	0.54 646	54 732	54 817	54 902	54 987	55 071	55 156	55 241	55 325	55 410
0.54	0.55 494	55 578	55 662	55 746	55 830	55 914	55 998	56 082	56 165	56 249
0.55	0.56 332	56 416	56 499	56 582	56 665	56 748	56 831	56 914	56 996	57 079
0.56	0.57 162	57 244	57 326	57 409	57 491	57 573	57 655	57 737	57 818	57 900
0.57	0.57 982	58 063	58 144	58 226	58 307	58 388	58 469	58 550	58 631	58 712
0.58	0.58 792	58 873	58 953	59 034	59 114	59 194	59 274	59 354	59 434	59 514
0.59	0.59 594	59 673	59 753	59 832	59 912	59 991	60 070	60 149	60 228	60 307
0.60	0.60 386	60 464	60 543	60 621	60 700	60 778	60 856	60 934	61 012	61 090
0.61	0.61 168	61 246	61 323	61 401	61 478	61 556	61 633	61 710	61 787	61 864
0.62	0.61 941	62 018	62 095	62 171	62 248	62 324	62 400	62 477	62 553	62 629
0.63	0.62 705	62 780	62 856	62 932	63 007	63 083	63 158	63 233	63 309	63 384
0.64	0.63 459	63 533	63 608	63 683	63 757	63 832	63 906	63 981	64 055	64 129
0.65	0.64 203	64 277	64 351	64 424	64 498	64 572	64 645	64 718	64 791	64 865
0.66	0.64 938	65 011	65 083	65 156	65 229	65 301	65 374	65 446	65 519	65 591
0.67	0.65 663	65 735	65 807	65 878	65 950	66 022	66 093	66 165	66 236	66 307
0.68	0.66 378	66 449	66 520	66 591	66 662	66 732	66 803	66 873	66 944	67 014
0.69	0.67 084	67 154	67 224	67 294	67 364	67 433	67 503	67 572	67 642	67 711
0.70	0.67 780	67 849	67 918	67 987	68 056	68 125	68 193	68 262	68 330	68 398
0.71	0.68 467	68 535	68 603	68 671	68 738	68 806	68 874	68 941	69 009	69 076
0.72	0.69 143	69 210	69 278	69 344	69 411	69 478	69 545	69 611	69 678	69 744
0.73	0.69 810	69 877	69 943	70 009	70 075	70 140	70 206	70 272	70 337	70 403
0.74	0.70 468	70 533	70 598	70 663	70 728	70 793	70 858	70 922	70 987	71 051
0.75	0.71 116	71 180	71 244	71 308	71 372	71 436	71 500	71 563	71 627	71 690
0.76	0.71 754	71 817	71 880	71 943	72 006	72 069	72 132	72 195	72 257	72 320
0.77	0.72 382	72 444	72 507	72 569	72 631	72 693	72 755	72 816	72 878	72 940
0.78	0.73 001	73 062	73 124	73 185	73 246	73 307	73 368	73 429	73 489	73 550
0.79	0.73 610	73 671	73 731	73 791	73 851	73 911	73 971	74 031	74 091	74 151
0.80	0.74 210	74 270	74 329	74 388	74 447	74 506	74 565	74 624	74 683	74 742
0.81	0.74 800	74 859	74 917	74 976	75 034	75 092	75 150	75 208	75 266	75 323
0.82	0.75 381	75 439	75 496	75 553	75 611	75 668	75 725	75 782	75 839	75 896
0.83	0.75 952	76 009	76 066	76 122	76 178	76 234	76 291	76 347	76 403	76 459
0.84	0.76 514	76 570	76 626	76 681	76 736	76 792	76 847	76 902	76 957	77 012
0.85	0.77 067	77 122	77 176	77 231	77 285	77 340	77 394	77 448	77 502	77 556
0.86	0.77 610	77 664	77 718	77 771	77 825	77 878	77 932	77 985	78 038	78 091
0.87	0.78 144	78 197	78 250	78 302	78 355	78 408	78 460	78 512	78 565	78 617
0.88	0.78 669	78 721	78 773	78 824	78 876	78 928	78 979	79 031	79 082	79 133
0.89	0.79 184	79 235	79 286	79 337	79 388	79 439	79 489	79 540	79 590	79 641
0.90	0.79 691	79 741	79 791	79 841	79 891	79 941	79 990	80 040	80 090	80 139
0.91	0.80 188	80 238	80 287	80 336	80 385	80 434	80 482	80 531	80 580	80 628
0.92	0.80 677	80 725	80 773	80 822	80 870	80 918	80 966	81 013	81 061	81 109
0.93	0.81 156	81 204	81 251	81 299	81 346	81 393	81 440	81 487	81 534	81 580
0.94	0.81 627	81 674	81 720	81 767	81 813	81 859	81 905	81 951	81 997	82 043
0.95	0.82 089	82 135	82 180	82 226	82 271	82 317	82 362	82 407	82 452	82 497
0.96	0.82 542	82 587	82 632	82 677	82 721	82 766	82 810	82 855	82 899	82 943
0.97	0.82 987	83 031	83 075	83 119	83 162	83 206	83 250	83 293	83 337	83 380
0.98	0.83 423	83 466	83 509	83 552	83 595	83 638	83 681	83 723	83 766	83 808
0.99	0.83 851	83 893	83 935	83 977	84 020	84 061	84 103	84 145	84 187	84 229

x	0	1	2	3	4	5	6	7	8	9
1.00	0.84 270	84 312	84 353	84 394	84 435	84 477	84 518	84 559	84 600	84 640
1.01	0.84 681	84 722	84 762	84 803	84 843	84 883	84 924	84 964	85 004	85 044
1.02	0.85 084	85 124	85 163	85 203	85 243	85 282	85 322	85 361	85 400	85 439
1.03	0.85 478	85 517	85 556	85 595	85 634	85 673	85 711	85 750	85 788	85 827
1.04	0.85 865	85 903	85 941	85 979	86 017	86 055	86 093	86 131	86 169	86 206
1.05	0.86 244	86 281	86 318	86 356	86 393	86 430	86 467	86 504	86 541	86 578
1.06	0.86 614	86 651	86 688	86 724	86 760	86 797	86 833	86 869	86 905	86 941
1.07	0.86 977	87 013	87 049	87 085	87 120	87 156	87 191	87 227	87 262	87 297
1.08	0.87 333	87 368	87 403	87 438	87 473	87 507	87 542	87 577	87 611	87 646
1.09	0.87 680	87 715	87 749	87 783	87 817	87 851	87 885	87 919	87 953	87 987
1.10	0.88 021	88 054	88 088	88 121	88 155	88 188	88 221	88 254	88 287	88 320
1.11	0.88 353	88 386	88 419	88 452	88 484	88 517	88 549	88 582	88 614	88 647
1.12	0.88 679	88 711	88 743	88 775	88 807	88 839	88 871	88 902	88 934	88 966
1.13	0.88 997	89 029	89 060	89 091	89 122	89 154	89 185	89 216	89 247	89 277
1.14	0.89 308	89 339	89 370	89 400	89 431	89 461	89 492	89 522	89 552	89 582
1.15	0.89 612	89 642	89 672	89 702	89 732	89 762	89 792	89 821	89 851	89 880
1.16	0.89 910	89 939	89 968	89 997	90 027	90 056	90 085	90 114	90 142	90 171
1.17	0.90 200	90 229	90 257	90 286	90 314	90 343	90 371	90 399	90 428	90 456
1.18	0.90 484	90 512	90 540	90 568	90 595	90 623	90 651	90 678	90 706	90 733
1.19	0.90 761	90 788	90 815	90 843	90 870	90 897	90 924	90 951	90 978	91 005
1.20	0.91 031	91 058	91 085	91 111	91 138	91 164	91 191	91 217	91 243	91 269
1.21	0.91 296	91 322	91 348	91 374	91 399	91 425	91 451	91 477	91 502	91 528
1.22	0.91 553	91 579	91 604	91 630	91 655	91 680	91 705	91 730	91 755	91 780
1.23	0.91 805	91 830	91 855	91 879	91 904	91 929	91 953	91 978	92 002	92 026
1.24	0.92 051	92 075	92 099	92 123	92 147	92 171	92 195	92 219	92 243	92 266
1.25	0.92 290	92 314	92 337	92 361	92 384	92 408	92 431	92 454	92 477	92 500
1.26	0.92 524	92 547	92 570	92 593	92 615	92 638	92 661	92 684	92 706	92 729
1.27	0.92 751	92 774	92 796	92 819	92 841	92 863	92 885	92 907	92 929	92 951
1.28	0.92 973	92 995	93 017	93 039	93 061	93 082	93 104	93 126	93 147	93 168
1.29	0.93 190	93 211	93 232	93 254	93 275	93 296	93 317	93 338	93 359	93 380
1.30	0.93 401	93 422	93 442	93 463	93 484	93 504	93 525	93 545	93 566	93 586
1.31	0.93 606	93 627	93 647	93 667	93 687	93 707	93 727	93 747	93 767	93 787
1.32	0.93 807	93 826	93 846	93 866	93 885	93 905	93 924	93 944	93 963	93 982
1.33	0.94 002	94 021	94 040	94 059	94 078	94 097	94 116	94 135	94 154	94 173
1.34	0.94 191	94 210	94 229	94 247	94 266	94 284	94 303	94 321	94 340	94 358
1.35	0.94 376	94 394	94 413	94 431	94 449	94 467	94 485	94 503	94 521	94 538
1.36	0.94 556	94 574	94 592	94 609	94 627	94 644	94 662	94 679	94 697	94 714
1.37	0.94 731	94 748	94 766	94 783	94 800	94 817	94 834	94 851	94 868	94 885
1.38	0.94 902	94 918	94 935	94 952	94 968	94 985	95 002	95 018	95 035	95 051
1.39	0.95 067	95 084	95 100	95 116	95 132	95 148	95 165	95 181	95 197	95 213
1.40	0.95 229	95 244	95 260	95 276	95 292	95 307	95 323	95 339	95 354	95 370
1.41	0.95 385	95 401	95 416	95 431	95 447	95 462	95 477	95 492	95 507	95 523
1.42	0.95 538	95 553	95 568	95 582	95 597	95 612	95 627	95 642	95 656	95 671
1.43	0.95 686	95 700	95 715	95 729	95 744	95 758	95 773	95 787	95 801	95 815
1.44	0.95 830	95 844	95 858	95 872	95 886	95 900	95 914	95 928	95 942	95 956
1.45	0.95 970	95 983	95 997	96 011	96 024	96 038	96 051	96 065	96 078	96 092
1.46	0.96 105	96 119	96 132	96 145	96 159	96 172	96 185	96 198	96 211	96 224
1.47	0.96 237	96 250	96 263	96 276	96 289	96 302	96 315	96 327	96 340	96 353
1.48	0.96 365	96 378	96 391	96 403	96 416	96 428	96 440	96 453	96 465	96 478
1.49	0.96 490	96 502	96 514	96 526	96 539	96 551	96 563	96 575	96 587	96 599

x	0	2	4	6	8
1.50	0.96 611	96 634	96 658	96 681	96 705
1.51	0.96 728	96 751	96 774	96 796	96 819
1.52	0.96 841	96 864	96 886	96 908	96 930
1.53	0.96 952	96 973	96 995	97 016	97 037
1.54	0.97 059	97 080	97 100	97 121	97 142
1.55	0.97 162	97 183	97 203	97 223	97 243
1.56	0.97 263	97 283	97 302	97 322	97 341
1.57	0.97 360	97 379	97 398	97 417	97 436
1.58	0.97 455	97 473	97 492	97 510	97 528
1.59	0.97 546	97 564	97 582	97 600	97 617
1.60	0.97 635	97 652	97 670	97 687	97 704
1.61	0.97 721	97 738	97 754	97 771	97 787
1.62	0.97 804	97 820	97 836	97 852	97 868
1.63	0.97 884	97 900	97 916	97 931	97 947
1.64	0.97 962	97 977	97 993	98 008	98 023
1.65	0.98 038	98 052	98 067	98 082	98 096
1.66	0.98 110	98 125	98 139	98 153	98 167
1.67	0.98 181	98 195	98 209	98 222	98 236
1.68	0.98 249	98 263	98 276	98 289	98 302
1.69	0.98 315	98 328	98 341	98 354	98 366
1.70	0.98 379	98 392	98 404	98 416	98 429
1.71	0.98 441	98 453	98 465	98 477	98 489
1.72	0.98 500	98 512	98 524	98 535	98 546
1.73	0.98 558	98 569	98 580	98 591	98 602
1.74	0.98 613	98 624	98 635	98 646	98 657
1.75	0.98 667	98 678	98 688	98 699	98 709
1.76	0.98 719	98 729	98 739	98 749	98 759
1.77	0.98 769	98 779	98 789	98 798	98 808
1.78	0.98 817	98 827	98 836	98 846	98 855
1.79	0.98 864	98 873	98 882	98 891	98 900
1.80	0.98 909	98 918	98 927	98 935	98 944
1.81	0.98 952	98 961	98 969	98 978	98 986
1.82	0.98 994	99 003	99 011	99 019	99 027
1.83	0.99 035	99 043	99 050	99 058	99 066
1.84	0.99 074	99 081	99 089	99 096	99 104
1.85	0.99 111	99 118	99 126	99 133	99 140
1.86	0.99 147	99 154	99 161	99 168	99 175
1.87	0.99 182	99 189	99 196	99 202	99 209
1.88	0.99 216	99 222	99 229	99 235	99 242
1.89	0.99 248	99 254	99 261	99 267	99 273
1.90	0.99 279	99 285	99 291	99 297	99 303
1.91	0.99 309	99 315	99 321	99 326	99 332
1.92	0.99 338	99 343	99 349	99 355	99 360
1.93	0.99 366	99 371	99 376	99 382	99 387
1.94	0.99 392	99 397	99 403	99 408	99 413
1.95	0.99 418	99 423	99 428	99 433	99 438
1.96	0.99 443	99 447	99 452	99 457	99 462
1.97	0.99 466	99 471	99 476	99 480	99 485
1.98	0.99 489	99 494	99 498	99 502	99 507
1.99	0.99 511	99 515	99 520	99 524	99 528
2.00	0.99 532	99 536	99 540	99 544	99 548

x	0	2	4	6	8
2.00	0.99 532	99 536	99 540	99 544	99 548
2.01	0.99 552	99 556	99 560	99 564	99 568
2.02	0.99 572	99 576	99 580	99 583	99 587
2.03	0.99 591	99 594	99 598	99 601	99 605
2.04	0.99 609	99 612	99 616	99 619	99 622
2.05	0.99 626	99 629	99 633	99 636	99 639
2.06	0.99 642	99 646	99 649	99 652	99 655
2.07	0.99 658	99 661	99 664	99 667	99 670
2.08	0.99 673	99 676	99 679	99 682	99 685
2.09	0.99 688	99 691	99 694	99 697	99 699
2.10	0.99 702	99 705	99 707	99 710	99 713
2.11	0.99 715	99 718	99 721	99 723	99 726
2.12	0.99 728	99 731	99 733	99 736	99 738
2.13	0.99 741	99 743	99 745	99 748	99 750
2.14	0.99 753	99 755	99 757	99 759	99 762
2.15	0.99 764	99 766	99 768	99 770	99 773
2.16	0.99 775	99 777	99 779	99 781	99 783
2.17	0.99 785	99 787	99 789	99 791	99 793
2.18	0.99 795	99 797	99 799	99 801	99 803
2.19	0.99 805	99 806	99 808	99 810	99 812
2.20	0.99 814	99 815	99 817	99 819	99 821
2.21	0.99 822	99 824	99 826	99 827	99 829
2.22	0.99 831	99 832	99 834	99 836	99 837
2.23	0.99 839	99 840	99 842	99 843	99 845
2.24	0.99 846	99 848	99 849	99 851	99 852
2.25	0.99 854	99 855	99 857	99 858	99 859
2.26	0.99 861	99 862	99 863	99 865	99 866
2.27	0.99 867	99 869	99 870	99 871	99 873
2.28	0.99 874	99 875	99 876	99 877	99 879
2.29	0.99 880	99 881	99 882	99 883	99 885
2.30	0.99 886	99 887	99 888	99 889	99 890
2.31	0.99 891	99 892	99 893	99 894	99 896
2.32	0.99 897	99 898	99 899	99 900	99 901
2.33	0.99 902	99 903	99 904	99 905	99 906
2.34	0.99 906	99 907	99 908	99 909	99 910
2.35	0.99 911	99 912	99 913	99 914	99 915
2.36	0.99 915	99 916	99 917	99 918	99 919
2.37	0.99 920	99 920	99 921	99 922	99 923
2.38	0.99 924	99 924	99 925	99 926	99 927
2.39	0.99 928	99 928	99 929	99 930	99 930
2.40	0.99 931	99 932	99 933	99 933	99 934
2.41	0.99 935	99 935	99 936	99 937	99 937
2.42	0.99 938	99 939	99 939	99 940	99 940
2.43	0.99 941	99 942	99 942	99 943	99 943
2.44	0.99 944	99 945	99 945	99 946	99 946
2.45	9.99 947	99 947	99 948	99 949	99 949
2.46	0.99 950	99 950	99 951	99 951	99 952
2.47	0.99 952	99 953	99 953	99 954	99 954
2.48	0.99 955	99 955	99 956	99 956	99 957
2.49	0.99 957	99 958	99 958	99 958	99 959
2.50	0.99 959	99 960	99 960	99 961	99 961

x	0	1	2	3	4	5	6	7	8	9
2.5	0.99 959	99 961	99 963	99 965	99 967	99 969	99 971	99 972	99 974	99 975
2.6	0.99 976	99 978	99 979	99 980	99 981	99 982	99 983	99 984	99 985	99 986
2.7	0.99 987	99 987	99 988	99 989	99 989	99 990	99 991	99 991	99 992	99 992
2.8	0.99 992	99 993	99 993	99 994	99 994	99 994	99 995	99 995	99 995	99 996
2.9	0.99 996	99 996	99 996	99 997	99 997	99 997	99 997	99 997	99 997	99 998
3.0	0.99 998	99 998	99 998	99 998	99 998	99 998	99 998	99 998	99 999	99 999

REFERENCES

[1] Driller, J. and Lizzi, F., "Therapeutic Application of Ultrasound: A Review", IEEE Engineering in Medicine and Biology Magazine, Vol. 6, No. 4, December 1987.

[2] Jones, R. V., *The Wizard War—British Scientific Intelligence, 1939–1945*, Coward, McCann, Geoghegan, New York, 1978.

[3] Papayannis, A. et al., "Multiwavelength Lidar for Ozone Measurements in the Troposphere and the Lower Stratosphere", Applied Optics, Vol. 29, No. 4, February 1990, p. 467.

[4] McDermid, S., Godin, S. M. and Lindquist, O., "Long-Term Measurements of Stratospheric Ozone", Applied Optics, Vol. 29, No. 25, September 1990, p. 3603.

[5] Ismail, S. and Browell, E. V., "Airborne and Spaceborne Lidar Measurements of Water Vapor Profiles: A Sensitivity Analysis", Applied Optics, Vol. 28, No. 17, September 1989, p. 3603.

[6] Urick, R. J., *Principles of Underwater Sound*, McGraw-Hill, New York, 1983.

[7] Officer, C. B., *Introduction to the Theory of Sound Transmission*, McGraw-Hill, New York, 1958.

[8] Elbaum, M. et al., "Maximum Angular Accuracy of Pulsed Laser Radar in Photocounting Limit", Applied Optics, Vol. 16, No. 7, July 1977, pp. 1982–1992.

[9] Teich, M. C. and Saleh, B. E. A., "Squeezed and Antibunched Light", Physics Today, June 1990.

[10] Nyquist, H., "Thermal Agitation of Electric Charge in Conductors", Physical Review, Vol. 32, July 1928, p. 110.

[11] Siegman, A. E., Proc. IEEE, Vol. 54, 1966, p. 1350.

[12] Elbaum, M. and Teich, M. C., Optics Communications, Vol. 27, No. 2, November 1978, 257.

[13] Gagliardi, R. M. and Karp, S., *Optical Communication*, Wiley, New York, 1976.

[14] Rice, S. O., "Mathematical Analysis of Random Noise", Bell System Technical Journal, Vols. 23 and 24, March 1944.

[15] Einstein, A. and Hopf, L. Ann. d. Physik, Vol. 33, 1910, pp. 1095–1115; Von Laue, M., Ann. d. Physik, Vol. 47, 1915, pp. 853–878; Einstein, A., Ann. d. Physik, Vol. 47, 1915, pp. 879–885; Von Laue, M., Ann. d. Physik, Vol. 48, 1915, pp. 668–680.

[16] Viterbi, A. J., *Principles of Coherent Communications*, Appendix A, McGraw-Hill, New York, 1966.

[17] Oppenheim, A. V. and Schafer, R. W., *Discrete-Time Signal Processing*, Prentice Hall, Englewood Cliffs, NJ, 1989.

[18] North, D. O., "Analysis of the Factors which Determine Signal/Noise Discrimination in Radar", Report PTR-6C, RCA Laboratories, June, 1943.

[19] Papoulis, A., *Signal Analysis*, McGraw-Hill, New York, 1977.

[20] Goldman, S., *Frequency Analysis, Modulation and Noise*, Dover Publications, New York.

[21] Whalen, A. D., *Detection of Signals in Noise*, Academic Press, New York, 1971, Chapters 6 and 7.

[22] Marcum, J. J., "A Statistical Theory of Target Detection by Pulsed Radar", RAND Research memos, RM-753 and RM-754 1947 and 1948 respectively; also available as Marcum, J. J. and Swerling, P., "Studies of Target Detection by Pulsed Radar", special monograph issue, IRE Trans. Information Theory, Vol. 6, April 1960.

[23] Lawson, J. L. and Uhlenbeck, G. E., *Threshold Signals*, Chapter 7, Antique Radio Classified, P.O. Box 2, Carlisle, Maine; originally Vol. 24 MIT Rad. Lab. Series.

[24] Emslie, A. G., "Coherent Integration", MIT Radiation Laboratory Report 103-5, 16 May 1944.

[25] Gabor, D. J., Inst. Electrical Engineers pt III, Vol. 93, p. 429.

[26] Woodward, P. M., *Probability and Information Theory with Application to Radar*, originally published by Pergamon Press, New York, 1953, now available only through University Microfilm at the University of Michigan, Ann Arbor, Michigan.

[27] Jones, R. V., *The Wizard War—British Scientific Intelligence, 1939–1945*, Coward, McCann, Geoghegan, New York, 1978, p. 124.

[28] Cramér, H., *Mathematical Methods of Statistics*, Princeton University Press, Princeton, NJ, 1963.

[29] Chang, C. B. and Tabaczynski, "Application of State Estimation to Target Tracking", IEEE Trans. on Automatic Control, Vol. 29, No. 2, February 1984, pp. 98–109.

[30] Klauder, J. R., Price, A. C., Darlington, S. and Albersheim, W. J., "The Theory and Design of Chirp Radar", Bell System Technical Journal, Vol. 39, July 1960.

[31] Cook, C. E. and Bernfeld, M., *Radar Signals*, Academic Press, New York, 1967.

[32] Nayfeh, A. H., *Introduction to Perturbation Techniques*, Wiley, New York, 1981.

[33] Thor, R. C., "A Large Time-Bandwidth Product Pulse-Compression Technique", IRE Transaction on Military Electronics, April 1962, p. 169.

[34] Dixon, R. C., *Spread Spectrum Systems*, Wiley, New York, 1984.

[35] Labianca, F. M., "Normal Modes, Virtual Modes, and Alternative Representations in the Theory of Surface-Duct Sound Propagation", Journal of the Acoustic Society of America, Vol. 53, No. 4, 1973, pp. 1137–1147.

[36] Wiener, N., *Extrapolation Interpolation and Smoothing of Stationary Time Series*, Wiley, New York, 1957.

INDEX

Aliasing, 62–64
in bandpass signals, 71, 72
in frequency, 62, 63
and sampling rate, 63, 64
in time, 82
Ambient Noise:
in oceans, 93
in optical systems, 32, 97
in radar and sonar, 32
Ambiguity, 11, 178, 183–190
doppler, 178, 179, 183
range, 178, 183–185
Ambiguity Function, 11, 183–190
footprint for LFM and HFM, 205–207
for Gaussian pulses, 188–190
generalized, 187
for large BT waveforms, comparison of LFM and HFM Signals, 205, 207
and Range-Doppler Resolution, 187
Analog to Digital (A/D) Converter, 10, 33
Analytic Signal, 74–77
and band limited assumption, 152
matched filter formulation for, 148, 149
and Parseval's Theorem, 77
representation of noise, 151
Antenna, 8, 218–224. *See also* Arrays for Radar and Sonar
3-dB beamwidth of, 220
beampatterns for rectangular and circular apertures, 220
gain of, 223, 224
impedance, conjugate matched, 35
and reciprocity of transmit and receive beampatterns, 221
temperature and thermal equilibrium with background, 32
Anti-Aliasing Filtering, 63, 67, 68
Approximation to Gaussian Distribution, *see* Gaussian
Arrays for Radar and Sonar, 4, 69, 221
3-dB beamwidth of, 221, 223
beamforming for discrete-time signals, 69
gain in sonar and radar, 223, 224
and grating lobes, 221
ASDIC, 2
Audio Signal and Frequency, 10
Autocorrelation Function, 28–31. *See also* Correlation Function
Averages:
ensemble and time, 29
law of, 20

Ballistic Missile Defense, 4
Ballistic Missile Early Warning System (BMEWS), 4
Band Limited Functions and Signals, 61, 62, 72–74, 101
 and analytic signal, 152
 and approximations, 63, 175
 duration of, 63, 72–74
 spectrum of, 62
Bandpass Filter, 118, 123
 approximation to matched filter, 123
Bandpass Signals, 70, 71
 aliasing of, 71, 72
 frequency representation of, 44, 70, 71, 118
 and Nyquist rate, 71, 72
Bandwidth, 7
 of carrier and video pulses, 72, 85
 definitions of (e.g. 3-dB), 72, 73, 154, 155
 fundamental limit on range resolution, 192
 of low pass filter, 100
 and signal duration, 63, 72–74
Baseband Signal, 10
Bats and Sonar, 1
Battle of Britain, 3
Baye's Decision Criterion, 11, 88–90, 94, 96, 109, 110, 113, 116
 and false alarm probability, 96
Baye's Theorem, 16, 17, 26, 27, 88, 90, 161
Beam Splitter, 39
Bell Telephone Laboratories, 3
Bernoulli Trials, 13–15
Bessel Function, Modified of Order Zero, I_0, 47, 130
 as optimum noncoherent detection characteristic, 130
Binominal Theorem and Distribution, 13–15
Bit Error Rate (BER), 91
Black-Body, 32, 33
Boltzmann's Constant, 33, 55
British Radar Development, 2
Brownian Motion, 31

Carrier Frequency, Definitions of, 70, 154
Causality, Causal Systems, 78
Central Limit Theorem, 21, 42, 140
Chang, C. B., 170
Characteristic Function, 18, 19
 of binominal distribution, 23, 24
 of Gaussian distribution, 19
 of Poisson distribution, 23
Charge-Coupled Devices (CCD), 11
Chirp, *see* Linear Frequency Modulation (LFM)
Chi-Square Distribution, 140, 143, 217
Clutter, 6, 93, 228
Coded Waveforms, 207–210
Coherence, 9
 differential, 135, 139
Coherent (Predetection) Integration, 11, 129, 135–139
 and chi-square distribution, 143
 and improvement in signal to noise ratio, 135–139
 performance of, 142–145
 and requirements on oscillator stability, 135
Coherent Detection, 11, 117–129, 145
 and matched filtering, 124
Coherent Light Source, 9
Colored Noise, 93
Communication Systems (Binary, M-ary), 88, 90, 91
Conditional Probability Distributions, 17, 18, 88
 and expectation, 18
 a priori and *a posteriori*, 17, 88, 89, 93
Continuous-Time Signals, 61, 62, 77–79
 processing of, 77–81
Convolution, 78
 filtering of discrete-time signals, 79, 83
 theorem, for real and complex functions, 79, 186, 190
Correlation Detection, 102, 103, 106
Correlation Function, 28. *See also* Autocorrelation Function
 of band limited signals, 86
 for baseband noise, 156
 of discrete and continuous-time signals, 30
 ensemble and time, 28, 29
Cosine Rolloff Filter, 67, 85

Count Function $k(t)$, 97
Cramer–Rao Bound, 11, 159, 160, 164–169
  and coherent and noncoherent observations, 164, 165
  range of applicability, 165, 166
  when phase information is used, 166–168
Cross Correlation Function, 29, 59, 213

Dark Current in Optical Systems, 97, 99, 106
Death Ray, 2
Decision Criteria, 11, 90–93. *See also* Bayes, Maximum Likelihood and Neyman Pearson
  and relation to detection threshold, 92
Decision Regions, 89
Deflection Signal to Noise Ratio (DSNR), 131, 132, 138
Delay of Discrete-Time Signals, 69
Delta Function, *see* Dirac Delta Function
Detection, *see* Envelope and Square Law Detection
Detection Probability (P_d), 91
  for coherent integration, 143, 145
  and Gaussian distribution, 95, 96, 105
  for generalized matched filtering, 145
  for matched filter, coherent and noncoherent, 105, 126, 127
  for noncoherent detection of single pulse, 126, 127
  for noncoherent integration, 141, 145
  and Poisson distribution, 98, 99, 107
Detection of Signals in Noise, 87–116
  coherent, *see* Coherent Detection
  noncoherent, *see* Noncoherent Detection
Detectors, 9, 10. *See* Envelope and Square Law Detection
Differential Absorption Lidar (DIAL), 5
Dirac Delta Function, 37, 57, 77
  properties of, 77
  representations of, 37, 62, 73, 74, 77
Discrete-Time Signals, 10, 11, 61–86
  convolutional filtering of, 79, 83
  correlation detection and, 102
  delay of, 69
  FFT filtering of, 81–84
  Fourier transform of, 61, 62
Distant Early Warning (DEW) Line, 4
Distribution Function, *see* Specific Distributions
Doppler:
  frequency, 135, 181, 187, 192
  invariant waveforms, 192
  measurements, and range rate estimation, 174, 181, 205, 207
  processing, 135
  resolution, 180, 183
  velocity ambiguity, 178, 179, 183
  velocity resolution, 166, 183

E/N_0, 55, 102–105
Einstein, VonLaue, and Narrowband Noise Representation, 43. *See also* References
Emitters and Absorbers, 31
Emslie, A. C., 135
Envelope (Linear) Detection, 45–47, 120–122, 130, 131, 151
  equivalence to square law detection, 131, 215
  and Rayleigh and Rice distributions, 46, 47
Equipartition Theorem of Statistical Mechanics, 55
Ergodic Process, 30
Error Function (ERF), 96, 104. *See also* Appendix
Estimation, 158–175. *See also* Range Measurement/Estimation
  of acceleration, 175
  of position, 169–175
  of range, using matched filter, 163
  of velocity, 170–175
Estimator, 158
  efficient, 160
  lower bound on variance of, 159
  maximum likelihood, 161
  mean value of, 158, 159

Estimator (*Continued*)
sufficient, 160, 161
unbiased, 159
Exponential Distribution, 16

Fading, 6
False Alarm Probability (P_{fa}), 90–92
for coherent integration, 143, 145
and Gaussian distribution, 94–96, 104, 105
for generalized matched filtering, 145
and matched filter, 104, 105
for noncoherent detection of single pulse, 125–127
for noncoherent integration, 141, 145
and Poisson distribution, 98, 99, 107
specification of, 229, 230
Fast Fourier Transform, *see* FFT
Federal Communications Commission, 10
FFT Filtering, 81–85
of long duration signals, 83, 84
and requirements on number of points, 82, 83
speed of, 81
and time aliasing, 82
Filtering, General Design Considerations, 66
Fourier Transform, 18, 30
of continuous and discrete-time signals, 61, 62, 80–82
Fractional Bandwidth, 7
in radar, sonar, and laser radar, 43, 169
Fraunhofer Defraction Integral (Far Field Diffraction), 220
Frequency Resolution, 80, 81, 86, 166

Gaussian:
approximation for poisson and binominal distributions, 23–25, 97
channel, 11, 93, 99–106
distribution, 16, 26, 27, 42, 93–96
Gaussian Noise, 28, 42, 43
narrowband representation, 43
and parameter estimation, 148, 164–169
and signal detection, 93–96, 99–106, 113, 118
Gaussian Random Variable, 20, 93, 104
characteristic function, 19
maximum likelihood estimate of mean and variance, 161, 162, 176
Ghost Images, 6
Green's Function Representation of Linear Time Invariant Systems, 78, 79

Heterodyning, 9
and laser radar, 32, 39
Hilbert Transform, 75, 76, 86
and analytic signals, 75–77, 149, 212–215
of band limited amplitude modulated signals, 77, 86
Huygen's Principle, 218
Hydrophone, 8
Hyperbolic Frequency Modulation (HFM), 12, 193, 202–205
advantage over LFM, 204, 205, 207
ambiguity function of, 204, 205–207
Doppler invariance of, 204
Hypothesis Testing, 88–92, 102, 103, 118
and maximum likelihood estimation, 161

Impulse Response of Continuous and Discrete Linear Systems, 77–79
and causality, 78, 102
Infrared Frequencies, 1, 34
In-Phase (I) and Quadrature (Q):
channels, 120, 121, 135, 136, 213, 214
components, 119, 124, 125, 129, 132–134, 137, 138, 214
Instantaneous Frequency, Definition of, 193
Intercontinental Ballistic Missile (ICBM), 4
Intermediate Frequency (IF), 9
in optical systems, 40
Interpolation, 63–70

analogy to filtering, 65, 66
and anti-aliasing filtering, 67
dependence on sampling rate, 67, 68
edge effects, 65–69
function, 63–69
Ionosphere, 2, 3

Kalman, R., 170
Kirchoff, and Black-Body Radiation, 31
Kronicker Delta, 110

Lagrange Multiplier, 91, 92, 177
Large Time Bandwidth Waveforms, 72, 73, 191–211
Lasers, 5
Law of Large Numbers, 20, 42
Light or Laser Detection and Ranging (LADAR, LIDAR), 1, 5
Likelihood Function, 89, 158
and maximum likelihood estimation, 161, 162
for multiple independent observations, 158
with unknown phase, 119, 120
Likelihood Ratio and Likelihood Ratio Test for Signal Detection, 87–90, 93, 94, 111, 118–120
for Baye's, maximum likelihood and Neyman Pearson criterion, 90, 92, 111–113
decision regions, 89, 92
implementation of, 92
for noncoherent detection, 119
for poisson distribution, 98, 106, 107
Lincoln Laboratories, 3
Linear Detector, *see* Envelope Detector
Linear Frequency Modulation (LFM or chirp), 12, 193–202
bandwidth as function of BT, 195–197
degradation in range resolution with large value of BT, 202
Doppler invariant properties, 192, 198, 200–202
Range-Doppler ambiguity of, 200
and Range-Time Sidelobes, 194
response of matched filter, 194
Linear Systems (Impulse Response, Transfer Function, Greens Function, etc.), 77–79, 102
Local Oscillator, 9, 39
Log-Normal Distribution, 27
Lord Rayleigh, 157, 218
Lord Rutherford, 2

Marconi, G., 2
Marcum's Q-Function, 126
Matched Filter, 11, 99–106, 107–109
analytic signal formulation, 148, 149
and coherent detection, 117, 124
coherent and noncoherent, 11, 123, 124, 151, 152
difference in signal to noise ratio of, 126–129, 144
and detection and false-alarm probability, 104, 105, 145
for flat noise power spectrum, 109
and Gaussian distribution, 105
generalized, 135, 139, 145
impulse response of, 102, 103, 112, 114, 115
for LFM waveform, 194
and maximization of signal to noise ratio, 107–110
and maximum likelihood estimation of range, 163
Maximum Likelihood Decision Criterion, 11, 90, 92, 96, 109–111, 116
Maximum Likelihood Estimation, 161–164
application to parameter estimation, 162
and Gaussian random variables, 161, 162, 176
and matched filter for range estimation, 163
Mean Value:
of Random Process, 28, 29
of Random Variable, 14
Medical Ultrasonics, 2
Minimum Detectable Signal in Optical Receiver, 41

MIT Radiation Laboratory Series, 5
Mixer, 9
Multipath, 6

Neyman-Pearson Criterion, 11, 90–92, 96, 106, 111, 116, 120
  advantage over Baye's and maximum likelihood criteria, 92
Neyman-Pearson Lemma, 92
  and maximum probability of detection, 96
NIKE Radar/Missile Systems, 4
Noise, 6. *See also* Gaussian Noise, Shot Noise and White Noise
  ambient, *see* Ambient Noise
  bandwidth, effect on matched filter output, 105, 156, 157
  bandwidth, two-sided, 37
  baseband and bandpass, 43, 44
  effect on range estimation, 153–157
  and mean square fluctuations of observables, 107, 131
  narrowband representation of, 43
  at output of envelope detector, 46
  thermal, 33–35, 93. *See also* Gaussian Noise
  types of, 31–33
Noise Figure, 48–51
  of attenuator, 50
  of front end amplifier, 51
  total for cascaded amplifiers connected by lossy elements, 49–51
Noise Power and Conjugate Matched Condition, 35
Noise Power Measurements, 51–55
  required integration time, 52
Noise and Random Processes, 28–60
Noise Temperature, 48–51
Noncoherent Detection of a Single Pulse, 117–129, 145
  comparison with coherent detection, 124–129
Noncoherent (Postdetection) Integration, 11, 129–134, 140–142, 145, 146
  and central limit theorem, 140
  and chi-square distribution, 140
  comparison of exact and approximate performance, 142
  improvement in signal to noise ratio, 139, 141
  performance of, 10, 141, 145
  and phase information, 132, 144
Noncoherent Operation, 9, 11, 117
Nonlinear Detectors, 9, 10, 120–124
Normal Distribution, *see* Gaussian Distribution
Nyquist, H., 33
Nyquist Rate, 63, 67, 69, 100, 102
  and bandpass signals, 71, 72, 118
Nyquist's Theorem, 32–34, 58

Optical:
  aperture, 8
  count function, 97
  heterodyne detection, 32, 39
  receiver, 6, 39–42, 96–99, 106, 107
    quantum efficiency of detector, 41
  systems, dark current and background radiation, 97
Overlap Save and Overlap-Add Methods, 83, 84
Ozone and Watervapor Profiles, 5

Parameter Estimation, 11, 148–177
  application to tracking and prediction, 169–175
  generalized, 158–159
Parseval's Theorem, 54, 60, 73, 77, 108
  practical version for finite limits, 54
Phase Measurement, 136, 137
Photodetector, 11, 39
Photoelectric Current, 96–97, 106
Photon:
  arrival times, 35, 36
  counting, 97–99, 106, 107
  energy, 6, 7, 42, 43
Pixels, 11
Planck Radiation Law, 33
Poisson Distribution, 15, 26, 97, 231
  approximation by Gaussian, 23–25, 42, 97
  as single parameter distribution, 42
Poisson Parameter, 15, 97
Poisson Shot Noise Channel, 11, 96–99, 106, 107

Poisson Statistics and Shot Noise, 42, 43
Post Detection Integration, *see* Noncoherent Integration
Power, Average, 129, 191, 229
Power-Aperture Product, 229
Power Spectral Density, 28–31
- of shot noise, 35–39
- for thermal noise, one-sided and two-sided, 35, 54

Probability, 13–27. *See also* Conditional Probability Distributions and Specific Distributions
- *a priori*, 3, 89
- density function (or distribution), 14
- detection, *see* Detection Probability
- false alarm, *see* False alarm probability
- and independent random variables, 17
- joint distribution, 17, 18

Propagation:
- attenuation, 218–220, 224–228
- in radar and sonar, 227
- random, 5, 6, 93

Pseudo Noise (PN), *see* Pseudo Random Noise Waveforms
Pseudo Random Noise (PRN) Waveforms, 207–210
Pulse Compression, 193, 194, 198
Pulse Repetition Frequency (PRF), 178
Pulse Train, Coherent and Noncoherent, 179–183

Quadrature Receiver, *see* Receiver, also In-Phase and Quadrature
Quantization Noise, 32
Quantum Limit in Optical Detectors, 39, 41

RADAR (Radio Detection and Ranging), 1–12, 87
- applications of, 3, 4, 87
- cross section, back scatter and bistatic, 225
- equation, 12, 224–227
- frequency nomenclature, 3
- Over The Horizon (OTH) radar, 3
- propagation loss, 227
- search problem, 228, 229
- synthetic aperture or side looking, 4

Radio Frequency (RF), 3, 9
Raman Scattering, 5
Random Process, 28–39, 93, 99–107
- for band limited functions, 86
- count function $k(t)$, 97
- definition of, 28
- as fundamental limit on observations, 5, 87, 88
- Gaussian, 99–105
- mean, 28
- Poisson, 106–107
- stationary, definition of, 28, 29

Random Variable, 14, 28, 88
- functions of, 25
- Gaussian, 16, 19, 20, 93–96, 104
- and independent observations, 100
- Poisson, 96–99

Range Measurement/Estimation, 15, 149–158
- ambiguity, 178
  - and resolution capability, 183–185
- and bandwidth, 157
- and coherent and noncoherent integration, 157
- for coherent or noncoherent matched filter, followed by square-law detector, 153, 157
- effects of noise, 153, 154
- and fractional bandwidth, 168
- for noncoherent observations, 164–166
- and phase information, 164, 166–169
- and signal to noise ratio, 157

Range Rate and Range Rate Estimation, 7, 171, 173–175, 181–183, 205
- by Doppler shift, 181, 205–207
- by tracking, 171–175

Range Resolution, 157, 166, 169, 172, 174, 183
Range-Time Sidelobes, 194
Rayleigh Distribution, 16, 17, 46, 47, 58, 112, 125
- and envelope detection, 45, 46
- and Rice distribution, 43

Receiver:
  complex, 213
  generic, 8
  quadrature, 120, 121, 135–137
Rectifier, 9, 10
Resolution, *see* Doppler, Frequency, Range, etc.
Reverberation, 6, 228
Rice Distribution, 16, 17, 46, 47, 58, 112, 125
  approximation by Gaussian distribution, 47, 58, 128
  and chi-square distribution, 143, 217
Riemann-Lebesgue Lemma, 74, 75, 85

Sampling Theorem and Oversampling, 61–69
  for carrier waveforms, 70–72
  and delay of discrete-time signals, 69
  and interpolation, 63, 64
Schwarz Inequality, 73, 108, 153, 160, 184
Scintillation, 6
Search Radar Equation, 228, 229
Shot Noise, 7, 32, 35–43, 93, 96–99, 106, 107
  channel, 106
  detection threshold, 98, 99, 106, 107
  fluctuations, smoothing of, 39, 43
  insignificance of in radar systems, 7, 43
  and Neyman-Pearson criterion, 106
  and Poisson statistics, 36, 42
  power spectral density, 35–39
  and quantum limit, 41
  and ratio of standard deviation to mean, 39
  signal to noise ratio, 43
Signal Duration and Bandwidth, 72–74
Signal to Noise Ratio (SNR), 43, 54, 108
  for coherent and noncoherent integration, 134, 139
  deflection, 131, 132
  integrated, 54
  for shot noise, 43
Single Parameter Distribution, 16, 23, 42
  ratio of standard deviation to mean, 16, 42
  and shot noise, 39
SONAR (Sonar Sound Detection and Ranging), 2
  applications of, 2, 87
  and bats, 1
  equation, 224, 227
  and medical applications, 2, 3
  propagation, 227
Speckle, 6
Square Law Detection, 120–122, 130–145, 151–156, 213–215
Squeezed States, 7
Standard Deviation, 14
Stationary Phase Principle, 197
Statistical Decision Theory, 87, 88

Tabaczynsky, J. A., 170
Target Strength in Sonar, 225, 227
Thermal Noise, 33–35. *See also* Gaussian Noise and White Noise
  power and matched impedance conditions, 35
  power spectral density, 33, 35
  reduction in optical receivers, 39
Time Bandwidth Product, Bounds on, 72–74. *See also* Large Time Bandwidth Product Waveforms
Tracking, 6, 169–175. *See also* Estimation of Position and Range Estimation
Transfer Function, 79
Transmission Loss, 227
Type 1 Error, 90
Type 2 Error, 90

Unit Frequency Step Function, 74
Unit Step Function, 110

Variance, 14
  maximum-likelihood estimate of, 162
v/c in Radar and Sonar, 7, 12, 202–207

Velocity, Orbital (Escape), 7, 182
measurements by tracking, 170, 171, 174, 175
Video Signal, 10
Viterbi, A., 43

Waveform Analysis, 178–183
White Noise, 33–35, 54–57, 99, 100, 104, 108, 109, 118, 124, 212. *See also* Thermal Noise and Gaussian Noise
band limited, 99, 100, 156
correlation function, 60, 100, 104, 124, 212
and Nyquist's theorem, 33–35
power spectral density, 33–35, 38, 54–57, 99, 108, 118
and shot noise, 38
Wiener Khinchine Theorem, 30, 31, 57, 100
Woodward, P. M., 157, 183